IN COMMEMORATION OF THE WORK OF
THE EIGHT THOUSAND YALE MEN
WHO TOOK PART IN THE WORLD WAR
1914-1918

HOW AMERICA WENT TO WAR

THE GIANT HAND

THE ROAD TO FRANCE I.

THE ROAD TO FRANCE II.

THE ARMIES OF INDUSTRY I.

THE ARMIES OF INDUSTRY II.

DEMOBILIZATION

HOW AMERICA WENT TO WAR

AN ACCOUNT FROM OFFICIAL SOURCES OF THE NATION'S WAR ACTIVITIES

1917-1920

Near the End of the Long Trail

From a photograph by the Signal Corps

THE ROAD TO FRANCE

I.

THE TRANSPORTATION OF TROOPS AND MILITARY SUPPLIES 1917-1918

BY BENEDICT CROWELL
THE ASSISTANT SECRETARY OF WAR AND DIRECTOR OF MUNITIONS 1917-1920

AND ROBERT FORREST WILSON
FORMERLY CAPTAIN, UNITED STATES ARMY

ILLUSTRATED WITH PHOTOGRAPHS FROM THE COLLECTIONS OF THE WAR AND NAVY DEPARTMENTS

NEW HAVEN
YALE UNIVERSITY PRESS
LONDON · HUMPHREY MILFORD · OXFORD UNIVERSITY PRESS
MDCCCCXXI

CONTENTS

PART I—THE LAND

PART II—THE PORT

PART III—THE SEA

APPENDICES

ILLUSTRATIONS

MAPS

PREFACE

WHEN the guns and the ammunition, the airplanes, the motor trucks, the general equipment, and the food and clothing of the American Army in the World War stood ready on the loading platforms of American factories and filled the army warehouses, the problem of supplying the American Expeditionary Forces with their necessities was as yet by no means solved. Those materials had still to travel a route the sources of which touched every producing point within the United States, and of which the main artery crossed the Atlantic. This was a military supply situation of unprecedented difficulty. No nation had ever attempted to maintain a great army over such a distance, nor was a line of supply ever so beset with peril.

Yet, while the forges and shops of the land were fighting their war manufacturing battles to ultimate triumph, our national genius for transportation rose superior to conditions and wrought the saving miracle of the struggle. It carried to France the two million men of the American Expeditionary Forces, together with such munitions and supplies as the grand strategy dictated. That achievement will probably forever stand as America's most signal contribution to the cause of the Entente and its associated nations.

The weight of American manpower proved to be a decisive factor in the defeat of Germany and her allies. Every American expected this result ultimately, but few expected it in 1918. Hence, even while the ships were carrying the conquering host of American troops to France, the military transportation organization was already preparing for the effort which was to freight across the ocean in 1919, according to interallied plan, an irresistible weight of American guns, ammunition, and other war materials. To this goal was directed our whole war industry. Had the war continued for another six months, it is probable—nay, certain—that the Atlantic would have buoyed up an eastward movement of American munitions every bit as astonishing as that transatlantic procession of Yankee troopships in the spring, summer, and early autumn of 1918.

Broadly analyzed, the supply of the American expedition fell into two divisions, *personnel* and *matériel*, men and things. As the War Department expanded in organization, each of these two divisions tended to segregate its activities from those of the other. There occurred a crystallization under pressure, the pressure of the emergency. Eventually all the activities relating to *personnel*—the conscription, classification, and training of troops, and the erection of military units—clustered within the administrative province of the Secretary of War and the General Staff. The enterprises in *matériel*—including principally the production of munitions—came to be the charge of the Assistant Secretary of War, who later bore the added title of Director of Munitions.

Such a demarcation was not sharply evident during the period of hostilities, although it actually existed. Nominally, the General Staff was in control of the production of supplies, because the Division of Purchase, Storage, and Traffic was a division of the General Staff; but this control was *only* nominal. In practice the General Staff Division of Purchase, Storage, and Traffic was the "overhead" through which the Director of Munitions functioned in the procurement and delivery of *matériel*. The allocation of supplies, once they had been manufactured and delivered to French ports, was properly the concern of the General Staff; but the manufacture of those supplies and their transportation to the point of delivery were not inherently general staff functions. It was hard for some general staff officers to see this distinction. The Director of Munitions produced the supplies and delivered them to the General Staff at the ports in France, functioning through the Division of Purchase, Storage, and Traffic. That great industrial organization was joined to the General Staff only because there seemed to be no other place for it in the administrative scheme as it then existed.

Military transportation was not a supply, but an agency: yet it was industrial in implication, and therefore, quite naturally, it became integral with the group of material activities directed by the Assistant Secretary of War.

Although the transportation organization's chief problems were those connected with the movement of military freight, it also had charge of the travel of troops. Thus it occurred that the office of the Assistant Secretary of War, otherwise concerned only with the inanimate elements of warfare, found itself dealing intimately with one of the most interesting, and certainly the most remarkable, of its human phases—the transportation of the men of the Army. The chapters which follow, therefore, while not slighting the less dramatic movement of our army supplies by land and sea, are devoted principally to the progress of the Expeditionary Army from its two million American homes to the shore of France.

The investigation from which this record results was conducted while the return of our troops from Europe was at its maximum volume and the transportation organization still intact. The story here presented comes not only from the official documents and files, but also from the memories of the men who did the work.

B. C. & R. F.-W.

Washington, D. C.,
November, 1920.

THE ROAD TO FRANCE

PART ONE
THE LAND

Photo by Harris & Ewing

GEORGE HODGES

DIED, MARCH 14, 1919

As head of the troop-movement office, he was field marshal of troop travel in America

CHAPTER I

A CONTRAST

IN the twenty-year period which separated our two overseas wars, that with Spain in 1898 and that with Germany in 1917-1918, America had made military progress; but in no province of the profession of war had the strides been longer or the distance covered greater than in the science of military transportation. The marching power of an army is almost, if not quite, as important to the success of its campaigns as its courage and fighting ability. Some of the greatest soldiers in all history—Hannibal, Julius Cæsar, Gustavus Adolphus, even Napoleon—built their military reputations largely upon the ability to move their troops. In their days, however, armies usually traveled on their own feet, and the great strategist staked as heavily upon the leg muscles of his soldiers as upon their spirit and valor. To-day, the motor truck, the passenger and freight train, and the ocean transport have been substituted for the thews and sinews of troops; and the movement of great bodies of soldiery has become a matter of the organization and management of the most intricate of all industries—transportation.

Our experiences in sending an overseas expedition to Cuba in 1898 showed the United States her shortcomings in military transportation and the magnitude of the lessons she had yet to learn. Thereafter our country maintained a force in the Philippine Islands and in other outlying possessions; and the necessity of providing replacements and supplies for these troops also gave our military authorities a measure of practice in the management of ocean transportation. A third most practical experience was to come in 1916, when with notable efficiency we assembled on the Mexican border the greatest concentration of American troops since the Civil War. Yet the Spanish War and all of the succeeding martial episodes of our history, added together, scarcely afforded us an adequate discipline and preparation for the transportation crisis which this nation was to face when she went to war with Germany. In comparison with the accomplishment of sending 2,000,000 men and their supplies across the Atlantic Ocean to France, the whole troop movement of the Spanish War takes rank with only the more commonplace phases of transportation in 1918; the maintenance of forces in our island possessions becomes a mere incident of traffic; and the border mobilization which excited the whole country in 1916 was little more than a hint of what, in the spring and summer of 1918, was to become a weekly, almost a daily, procedure.

The reader, if he is fully to appraise the quality and merit of the system which handled our military transportation during the recent hostilities, to understand the difficulties measured and mastered before that system could be what it was, to estimate the prodigious sum of its accomplishments and appreciate the smooth perfection of its processes, must begin with a quickened memory of 1898, when we made our last previous great effort in military transportation. Against the melancholy background of the war with Spain, the history of our military transportation since 1917 is brilliant indeed.

Tampa, on the Gulf coast, was the chief port of embarkation for the American expedition sent to Cuba in 1898. For what occurred in Tampa let us accept the testimony of an illustrious eyewitness; one who, so far from holding a prejudice, characterizes in lenient terms, even if he does not condone, the failures of those momentous days. In *The Rough Riders*,* Theodore Roosevelt refers to the commander of troops in 1898 as having "positively unlimited opportunity for the display of individual initiative," and as being in no danger "of finding his faculties of self-help numbed by becoming a cog in a gigantic and smooth-running machine."

* *The Rough Riders;* copyright, 1899, by Charles Scribner's Sons. Quotations herein used by permission of the publishers.

"If such a battalion chief wants to get anything or go anywhere he must do it by exercising every pound of resource, inventiveness, and audacity he possesses. The help, advice, and superintendence he gets from the outside will be of the most general, not to say superficial, character. . . . When he wishes to embark his regiment, he will have to fight for his railway cars exactly as he fights for his transport when it comes to going across the sea; and on his journey his men will or will not have food, and his horses will or will not have water and hay, and the trains will or will not make connections, in exact correspondence to the energy and success of his own efforts to keep things moving straight."

The Rough Riders had been recruited among adventurous millionaires, clubmen, and college athletes of the East, and among the rough cowmen, sheriffs, "lungers," and prospectors of the Far West, with a spirited and eager addition from the other principal sections of the United States. Leonard Wood was colonel of the regiment and Theodore Roosevelt lieutenant colonel. The organization gathered at San Antonio, Texas, and, after two or three weeks of military training, received orders to proceed by train to Tampa. The regiment entrained in seven sections, Colonel Wood commanding the first three and Lieutenant Colonel Roosevelt the remaining four. It took the transportation service of that time four days to move the sections from San Antonio to Tampa, "and," comments Colonel Roosevelt in *The Rough Riders*, "I doubt if anybody who was on the trip will soon forget it." The railroads had promised to move the regiment to Tampa in forty-eight hours. "Our experience in loading was enough to show that the promise would not be made good. There were no proper facilities for getting the horses on or off the cars, nor for feeding or watering them; and there was endless confusion and delay among the railway officials. I marched my four sections over in the afternoon, the first three having taken the entire day to get off. We occupied the night."

After describing some of the difficulties in entraining his troops and their impedimenta, Colonel Roosevelt continued:

"Meanwhile I superintended not merely my own men, but the railroad men; and when the delays of the latter, and their inability to understand what was necessary, grew past bearing, I took charge of the trains myself, so as to ensure the horse cars of each section being coupled with the baggage cars of that section." But after the animals and baggage were successfully loaded, it was discovered that no passenger cars were in evidence. It was then nearly midnight. Some of the men had scattered in the darkness, and it was necessary to gather them up again from the "vile drinking-booths around the stock yards." Finally this was accomplished, and, the passenger cars still not having come, the men were ordered to lie down beside the railroad tracks. Not until dawn did the passenger coaches arrive.

During four hot and dusty days the train crawled across the South. Yet the officers of the Rough Riders did not find time dragging on their hands. "There was enough delay and failure to make connections on the part of the railroad people to keep me entirely busy. . . . It happened that we usually made our longest stops at night, and this meant we were up all night long. . . .

"It was four days later that we disembarked, in a perfect welter of confusion. Tampa lay in the pine-covered sand-flats at the end of a one-track railroad, and everything connected with both military and railroad matters was in an almost inextricable tangle. There was no one to meet us or to tell us where we were to camp, and no one to issue us food for the first twenty-four hours; while the railroad people unloaded us wherever they pleased, or rather wherever the jam of all kinds of trains rendered it possible. We had to buy the men food out of our own pockets, and to seize wagons in order to get our spare baggage taken to the camping ground which we at last found had been allotted to us."

Eventually, however, the camp was made. Military drilling began. The war correspondents—Richard Harding Davis and Frederic Remington among them—called; social life began; and (a circumstance which, for all we know, may have had a

Photo by Brown Bros.

ROUGH RIDERS' CAMP AT TAMPA
(1898)

Copyright by Underwood & Underwood, N. Y.

THE ROUGH RIDERS WAITING TO EMBARK FOR CUBA

Photo by Brown Bros.

"A PERFECT WELTER OF CONFUSION"

Photo by Brown Bros.

TRANSPORTS AND DOCK AT TAMPA
(1898)

bearing on events nineteen years later) the military *attachés*, including the German *attaché*, came to Tampa to look on and, no doubt, privately to marvel.

Four or five days later, just as the Rough Riders were preparing for an indefinite stay, notification came that the expedition would start at once for a destination unknown, and that of the Rough-Rider Regiment eight troops without their horses would be taken along.

"It was the evening of June 7 when we suddenly received orders that the expedition was to start from Port Tampa, nine miles distant by rail, at daybreak the following morning; and that if we were not aboard our transport by that time we could not go. We had no intention of getting left, and prepared at once for the scramble which was evidently about to take place. As the number and capacity of the transports were known, or ought to have been known, and as the number and size of the regiments to go were also known, the task of allotting each regiment or fraction of a regiment to its proper transport, and arranging that the regiments and the transports should meet in due order at the dock, ought not to have been difficult. However, no arrangements were made in advance; and we were allowed to shove and hustle for ourselves as best we could, on much the same principles that had governed our preparations hitherto.

"We were ordered to be at a certain track with all our baggage at midnight, there to take a train for Port Tampa. At the appointed time we turned up, but the train did not. The men slept heavily, while Wood and I and various other officers wandered about in search of information which no one could give. We now and then came across a Brigadier General, or even a Major General; but nobody knew anything. Some regiments got aboard the trains and some did not, but as none of the trains started this made little difference. At three o'clock we received orders to march over to an entirely different track, and away we went. No train appeared on this track either; but at six o'clock some coal-cars came by, and these we seized. By various arguments we persuaded the engineer in charge of the train to back us down the nine miles to Port Tampa, where we arrived covered with coal-dust, but with all our belongings.

"The railroad tracks ran out on the quay, and the transports, which had been anchored in midstream, were gradually being brought up alongside the quay and loaded. The trains were unloading wherever they happened to be, no attention whatever being paid to the possible position of the transport on which the soldiers were to go. Colonel Wood and I jumped off and started on a hunt, which soon convinced us that we had our work cut out if we were to get a transport at all. From the highest General down, nobody could tell us where to go to find out what transport we were to have. At last we were informed that we were to hunt up the depot quartermaster, Colonel Humphrey. We found his office, where his assistant informed us that he didn't know where the Colonel was, but believed him to be asleep upon one of the transports. This seemed odd at such a time; but so many of the methods in vogue were odd, that we were quite prepared to accept it as a fact. However, it proved not to be such, but for an hour Colonel Humphrey might just as well have been asleep, as nobody knew where he was and nobody could find him, and the quay was crammed with some ten thousand men, most of whom were working at cross purposes.

"At last, however, after over an hour's industrious and rapid search through this swarming ant-heap of humanity, Wood and I, who had separated, found Colonel Humphrey at nearly the same time and were allotted a transport—the *Yucatan*. She was out in midstream, so Wood seized a stray launch and boarded her. At the same time I happened to find out that she had previously been allotted to two other regiments—the Second Regular Infantry and the Seventy-first New York Volunteers, which latter regiment alone contained more men than could be put aboard her. Accordingly, I ran at full speed to our train; and leaving a strong guard with the baggage, I double-quicked the rest of the regiment up to the boat, just in time to board her as she came into the quay, and then hold her against the Second Regulars and the Seventy-first, who had arrived a little too late, being a shade less ready than we were in the matter of individual initiative. There was a good deal of expostulation, but we had possession; and as the ship could not contain half of the men who had been told to go aboard her, the Seventy-first went away, as did all but four companies of the Second. These latter we took aboard. Meanwhile a General had caused our train to be unloaded at the end of the quay furthest from where the ship was; and the hungry, tired men spent most of the day in the labor of bringing down their baggage and the food and ammunition."

Loading accomplished, the *Yucatan* dropped down the stream and anchored; and because of a confusion in orders, the whole expedition remained on its transports in Tampa Bay for nearly a week thereafter. The soldiers, packed into the ships like sardines, stewed and sweltered in the burning heat of a subtropical June.

For a contrast to this distressing picture, let us anticipate a noteworthy episode of this narrative by following the train movement and overseas embarkation of the infantry regiments of the Seventy-ninth Division in the recent war. The Seventy-ninth was a division of the National Army, made up largely of drafted troops and formed and trained at Camp Meade, Maryland, near the city of Washington. It was called to France early in July, 1918, when the paramount need of the American Expeditionary Forces was for infantry. Consequently this division moved without its artillery.

The preparation for the embarkation of the Seventy-ninth began on the nation's Independence Day, when three freight trains moved out of Camp Meade, bearing the division's baggage directly to the ship's side at Hoboken. These trains were followed on July 5 by four passenger sections carrying the headquarters company of the division and detachments to go aboard the transports in advance of the main body of men, there to be instructed in the correct assignment of men to their quarters and in the routine which must be followed while the division was crossing the ocean.

The complete movement of the infantry units of the division, numbering over 18,000 men, or thirty-six times as many as Colonels Wood and Roosevelt had led aboard the *Yucatan*, was accomplished in the two following days, July 6 and 7. Two railroads connect Camp Meade with the water front at New York, the Pennsylvania and the Baltimore & Ohio. Not only were the 18,000 men handled over these two lines in two days, but the movement was accomplished in less than an eleven-hour period in each of the two days, all the trains being put through from start to destination each day between three o'clock in the afternoon and two o'clock in the early morning. Loaded trains left the camp at intervals as brief as fifteen minutes.

On July 6 the first train started from Camp Meade at 4.00 p.m. with 343 men aboard, and the last of the sixteen sections which departed on this day pulled out at 8.45 p.m. with 549 men on board. One of the July 6 trains carried 756 men, and the trains averaged more than 500 passengers each. The average time of loading a train was less than twenty minutes. Also on July 6, a heavy consignment of the division's baggage left Camp Meade by freight for Newport News, Virginia, whence it was to cross the ocean in a cargo convoy and rejoin the division in France.

On the following day, July 7, the movement was even heavier, for nineteen passenger trains left Camp Meade with an average of more than 500 men each. The first section left the camp at 3.15 p.m., and the last section of the regular movement at 8.30 o'clock—a rate of departure requiring each section to be loaded in about twenty minutes, average time. Two supplementary sections, following later in the evening, carried the soldiers detailed to see that neither men nor materials were left behind and that the camp was ready to be turned over in good order to the division's successors.

It should be noted that these troops went directly from their training camp to their ships, without any delay at an embarkation camp. The troop trains began arriving at Hoboken in the early evening of each of the two days; and

Copyright by Underwood & Underwood, N. Y.

THE ROUGH RIDERS SEIZE THE *YUCATAN*

Copyright by Underwood & Underwood, N. Y.

THE ROUGH RIDERS EMBARK

From The War College Collection

TROOPS ENTRAINING AT CAMP MEADE
(1918)

Photo by Signal Corps

A DETRAINMENT AT PORT OF EMBARKATION

after (1) a quick and efficient inspection, (2) the issue, if necessary, of articles of clothing or ordnance to make up the full personal equipment of each soldier, and (3) the final medical and other necessary inspections, the men went directly aboard their ships. The trains reached Hoboken every few minutes, but so efficiently was the embarkation handled that there was no cramming at the piers, and there was always transport space ready for each unit as it arrived. When a transport was loaded, it dropped down to the Lower Bay and anchored. Early on the morning of the third day the whole convoy was ready to sail.

Note, too, that, although this movement happens to have been the quickest transfer of a division during the war, it was not then regarded by the military transportation organization as anything out of the ordinary. It was simply one incident in the routine. Not until the armistice gave leisure for retrospection did the service check up its figures and discover that it had set a record.

In Germany, in France, and in England, too, the regular railroad service to civilians was greatly upset and, in periods of heaviest military traffic, sometimes suspended altogether, because the rail facilities in those countries were burdened beyond their limits by the necessities of the armies. The meager passenger train equipment of the Continental railroads was utterly inadequate to military needs, and troops invariably traveled in accommodations normally assigned to live-stock and other freight. The French box car with its war-time loading instructions, "40 *hommes*, 8 *chevaux*,"—40 men or 8 horses,—the "bucko special"—was one of the most familiar jokes of the war to the derisive Yanks, who at home had ridden in accommodations which even peace-time Europe would regard as *de luxe*—sleeping cars whenever the journey was of twenty-four hours' duration or longer, and in all other circumstances comfortable coaches or even, sometimes, parlor cars.

America, during the war period which ended with the armistice, transported over her rails nearly 9,000,000 soldiers with absolutely no disorganization of regular service, and with only such curtailment of some of the luxuries of inland travel as national economy demanded. On the day when the nineteen troop trains of the Seventy-ninth Division moved away from Camp Meade, and right in the midst of the movement, the "Congressional Limited," the crack passenger train between Washington and New York, left Washington as usual and on schedule. Perhaps not one of the passengers on that train realized that he was going fifty or sixty miles an hour in the midst of an ocean-bound division of Yankee troops, all traveling as comfortably, as swiftly, and as safely as he.

And what was happening on the rails between Camp Meade and Hoboken during those two days was being duplicated in almost every section of the country. The average number of troops transported by rail in the United States during the month of July, 1918, was over 35,000 a day; so that the movement from Camp Meade on the heavier of the two days did not amount to more than one-quarter of the total volume of inland military passenger traffic on that day.

The circumstances which, at Tampa, Colonel Roosevelt ironically characterized as "odd," did not exist in the transportation of American troops during the recent war. If they had existed, General Pershing's force of 2,000,000 men never would have existed. In Tampa, fewer soldiers were mobilized altogether than were transported from Camp Meade and loaded aboard ship in two days. Had the oddities of Tampa been repeated at such a port as New York in 1918, who can picture the indescribable result, the confusion, the bitter disgrace to America, the almost certain disaster to our cause? Men were being moved to France at the rate of 10,000 a day—more than a division every three days, more in a week than mobilized at Tampa altogether. With the embarkation camps at New York almost continuously packed with troops to utmost capacity, with other tens of thousands always on the rails moving steadily and inexorably toward the Port, and with System there in "the neck of the bottle" so ordering the operation that the ships snatched away their loads at precisely the rate of flow of the olive-drab torrent to the sea, what chaos, what an inextricable tangle, had the system faltered!

But, as one military bureau* expressed its astonishment, "the almost unbelievable facts" were true "that there was never any serious congestion at camp or ports, never any serious delays or accidents *en route*, and that no troopship was delayed by lack of troops or sailed without its full capacity of troops, except . . . on account of the influenza epidemic."

The 1918 commander of overseas troops had no need to display that "individual initiative" of Spanish War days, or to fight for his equipment, his railroad cars, or his transportation. Nor did he need to fear, with M. Demolins, whose book occupied Colonel Roosevelt on the railway trip from San Antonio to Tampa, that his individual faculties might become atrophied in consequence of his being merely "a cog in a vast and perfectly ordered machine." If he lost individuality, it would not be on that account. The vast and perfectly ordered machine existed in 1918, but the combatant officer was not even a cog in it. He was merely a passenger transported by it. The whole affair of his travel from training camp to debarkation port in France was handled for him by an expert organization which attended to every detail with intimacy and solicitude such as the most timorous and inexperienced of Cook's tourists never knew. The commander of combatant troops received orders to proceed to France, and thereafter his concern was merely to maintain discipline among his men during the journey. Otherwise he had little to do but fold his hands and await arrival in France.

The transportation organization told him when to move and how to prepare for the journey. When he marched his troops to the camp and railroad station, he found the trains ready for him. Transportation officers on the spot had arranged that. Transportation officers superintended the loading of the trains. Transportation officers routed the trains across the country, saw to it that they made their junction connections,

* Operations Branch, Operations Division, General Staff. Résumé of Activities, July 1, 1918–June 30, 1919.

kept them up to schedule. Transportation officers handled the detraining at the embarkation camps at the seaboard and assigned the troops to quarters. The troop commander was not even charged with the responsibility of seeing that his men were properly equipped. Experts of the transportation organization, conversant with the latest of the ever-changing orders and regulations, attended to this, confiscating excess and unauthorized articles, issuing new if the equipment were worn or deteriorated, and finally placing each man aboard the transport with his outfit in good condition and exactly complete, no more and no less. Troops on the trip by rail had to provide their own food, but the cooking facilities, kitchen cars and the like, were supplied by the transport organization. The Service had the last word on what men should go to France and what ones should not, for the final weeding-out occurred at the Port of Embarkation. If the unit had—as it usually had—left stragglers behind at training camp or elsewhere, even the difficult duty of bringing up these men, no matter in what part of the country they were, rested not on the commander, but on the broader shoulders of the transportation organization, which maintained a traveling military police force for this purpose—a huge business in itself.

A vast and perfectly ordered machine! So vast it was that no one mind could encompass its intricacies, yet so perfectly ordered that the smoothness of its performance must sometimes have astonished even its creators. There was never a malfunction of crucial importance; never a grave slip-up. Its capacity for handling men seemed to have no limit. It delivered troops to France so fast that our own soldiers construed the published figures of American overseas transportation as exaggerations meant to hearten the Allies and assail the morale of the enemy. Not Germany, not France, not one of even the most militaristic nations of the world had ever accomplished anything like it. In a most exacting branch of the profession of war, peace-loving America showed herself to be not the neophyte, but the master.

Copyright by Underwood & Underwood, N. Y.

ON THE DOCK AT TAMPA

Photo copyright by International Film Service

HARVARD HOSPITAL UNIT
LEAVING BOSTON (May, 1917)

Photo from Brown Bros.

ROUGH RIDING FOR THE ROUGH RIDERS
(Tampa, 1898)

Copyright by Underwood & Underwood, N. Y.

MARINES ENTRAINING FOR EMBARKATION

CHAPTER II

THE START OF THE VANGUARD

AMERICAN troops began moving toward France without undue delay after the declaration of war. Having taken the plunge, America was at once to realize what it involved: fighting, our own men facing the enemy in Flanders, engagements, American casualty lists—and all of this soon. We had a Regular Army—small, as forces had come to be regarded; not thoroughly equipped in the ultramodern sense; but intrepid, highly trained, and efficiently commanded by the graduates of West Point. We had, moreover, some one hundred thousand National Guardsmen still bronzed and hard from service at the Mexican border—a force which merited the respect of any enemy because of its recent training. The first act of Congress after adoption of the war resolution was to introduce a selective service bill which was soon to result in the registration and mobilization of the entire manpower of the greatest of republics. The land began to roar with the hammer-blows of workmen building the first thirty-two training camps. All other building suddenly ceased, and the rails grew heavy with freight—principally building materials—for the new mobilization cities. Officers' training camps had sprung up in half a dozen centers. Men were beginning to disappear from offices, from factories, from college classrooms. The war was upon us; its inexorable process of consumption had begun. The great machine was in motion.

Almost at once the Allies startled us with appeals for men. We had not known how desperate was their plight. Missions, headed by men high in the administration of the struggle against Germany, began arriving in the United States. The scales fell from our eyes. If the impression existed that we could take a year or two for preparation and then send a moderate force abroad to add the finishing touch to a victory which the Allies in themselves had the power to achieve, that impression was quickly dispelled. The Allies were bleeding to death. Germany was winning, nothing less. The morale of the French, who thus far had successfully held the "frontier of freedom," was running out like water through a sieve. There had been mutinies—officers shot by their men. American soldiers in France, not merely to be to the people of Europe the visible token of our moral support and of the material aid that would come presently, but soldiers forthwith for the actual fighting; green soldiers if necessary, to be brigaded with the veteran troops of the Allies for speed in training; men, soldiers, Americans, in the utmost possible numbers, in the quickest possible time—such was the desperate need. Without such aid, the cause of civilization might go under; it might, even if the aid came.

Thus, from the start, a heavy responsibility rested upon the transportation organization. As it wrought for success or failure, so might the destiny of humanity turn.

General Pershing, summoned from the Rio Grande, set up in Washington the headquarters of a new element in the world struggle, the American Expeditionary Forces. The Government had already secured the first ships of its future transport fleet, and these were assembling in New York, which, for the sake of secrecy in our military movements, had lost its metropolitan identity in the public press and had become, noncommittally, "an Atlantic port." The British ship *Baltic* slipped out of port, and a few days later the cable astonished millions of Americans by announcing the arrival in England of General Pershing and his staff. The A. E. F. was a fact. Already orders had been issued; and the transportation organization, still no more than the embryo of what it was shortly to become, had begun to function. The First Division was beginning to assemble for the voyage across the Atlantic. America had inaugurated the offensive.

General Pershing, however, was not the first to wear the American uniform in Europe. Several hospital organizations had preceded him, the earliest care of the Government being to provide accommodations and treatment for the sick and wounded of the forthcoming expedition. To Base Hospital No. 4, organized at Lakeside Hospital in Cleveland, Ohio, under the leadership of the eminent surgeon, Dr. George W. Crile, now commissioned in the Medical Corps, fell the honor of being the first American military unit to reach Europe. This organization sailed on May 8, 1917, on the S. S. *Orduna* of the Cunard Line. It was composed of 34 officers, 156 enlisted men, 64 nurses, and 4 civilians, and it carried a complete hospital equipment. Landing in England, these soldiers gave Europe its first glimpse of the American khaki. Three days later the hospital organized at Harvard University, Cambridge, and called Base Hospital No. 5, sailed on the Cunarder *Saxonia*, to be followed the next day by Base Hospital No. 2, which had been organized at the Presbyterian Hospital in New York, and which sailed on the American liner *St. Louis*. On May 19, two other base hospitals, with their equipment, left America for France: the university unit of St. Louis, Missouri, on the S. S. *St. Paul*, American Line, and now called Base Hospital No. 10; and Base Hospital No. 12, the Northwestern University unit, of Evanston, Illinois, on the American liner *Mongolia*.

General Pershing sailed on May 28, nine days after the last of the pioneer five base hospitals. He went with the understanding that the First Expeditionary Division was to follow him after a brief interval. He was still in Washington when the first overseas orders to combat troops went out in the form of a telegram from Colonel William M. Cruikshank, adjutant general, to the commanding general of the Southern Department of the Army at Fort Sam Houston, Texas. These orders were as follows:

"1. The following organizations are designated for foreign service: Sixteenth, Eighteenth, Twenty-sixth, Twenty-eighth Infantry Regiments; headquarters and four motor truck companies, personnel only; outpost company from First Battalion Signal Corps; Motor Ambulance Company No. 6; Motor

Field Hospital No. 6; two motor-truck company machine shops with personnel complete.

"2. Use all resources of your department to organize infantry regiments and equip them accordingly. Fill up organizations to strength prescribed by transfer of enlisted men from other organizations or by recruits. Wire requests for shortages direct to bureau chief concerned, who will have shortages meet organization at port of embarkation. Commissioned personnel will be assigned by the War Department. Expedite organization and equipment in every way possible. Report by wire when organizations are ready to move by rail."

A day or so later a telegram was dispatched to the commander at Fort Sam Houston directing that the First Division be organized and ready to entrain by June 1. This telegram was signed by General McCain, the adjutant general. There is another historic document in the files at Washington on the letter-head of the A. E. F. Headquarters, Washington, D. C., dated May 24, requesting the adjutant general to substitute the outpost company of the Second Field Signal Battalion for that of the First Battalion named in the original order. This letter is signed by General Pershing.

On May 24 General Pershing instructed the commanding general at Fort Sam Houston to order the four infantry regiments of the First Division, the 16th, 18th, 26th, and 28th, to take their existing equipment with them, including their Benét-Mercier machine guns. The rest of a complete field equipment was to meet the regiments at the Port of Embarkation and be loaded on ships constituting part of the convoy. The system of having troops carry along their equipment wherever they went was rooted deep in the traditions of the American Army, though it was to be drastically changed before the fighting came to an end.

Meanwhile, a chief concern of the military line organization in Washington was the tactical reorganization of units for overseas duty. The old American regiment with its 1,000 or 1,200 men and companies of 100 members would not fit in with the scheme of organization adopted by both the British and the French: namely, companies of 264 men and regiments with an approximate strength of 2,800 officers and men. One of General Pershing's first duties in his foreign service was to study this problem and to recommend a plan for the reorganization of American troop units. Late in May the Government organized a mission of eleven army officers to go to France immediately and study the many problems connected with the landing and establishment of a great American force, among these problems being that of army organization. Meanwhile, upon the arrival of French officers with the mission headed by M. Viviani and Marshal Joffre, the Army War College, after consultation with these experts, drew up a table of organization for an American division. General Pershing approved this scheme of organization before he sailed and requested that similar plans be formulated to cover the organization of army corps and field armies.

Washington decided to send, in addition, a regiment of marines with the first convoy. On May 16 the Secretary of War wrote to the Secretary of the Navy requesting that this regiment be organized according to the new specifications, with a headquarters company, three battalions, and an aggregate strength of 2,779 officers and men. It was also requested that replacement troops to the number of 1,000 either accompany the marine regiment or follow it immediately. Meanwhile the four infantry regiments of the First Division were to be brought up to the newly authorized strength by transfer of men from other units and by recruiting.

Major General William L. Sibert, later Chief of the Chemical Warfare Service, was assigned by the Secretary of War to the command of the First A. E. F. Division. General Sibert at once went to New York, set up his headquarters, began collecting his personal staff, and made arrangements for the loading of the first convoy. In a telegram from General Sibert to the War Department we catch a glimpse of the stir of preparation for combat which animated the military service in these early days:

"Information received here indicates that various bureau

chiefs of the War Department are ordering personnel and supplies to New York for transportation within first convoy. They cannot be accommodated on vessels procured. Arrangements have been made for the transportation of units itemized on sheet accompanying your letter May 26, and in addition one field bakery company and stevedores. No room for any other. Provisions made for only one ambulance company and no more can be accommodated, neither personnel nor equipment. No room for motor vehicles, other than those mentioned in list referred to above. Cargo space available is now overtaxed, and some portions will have to be left behind unless additional ships can be secured."

This telegram shows the eagerness of the various branches of the Army to place their men and materials in France at the earliest possible date. This zeal, if it had remained unchecked by central authority, might have entailed disaster to the great overseas movement or, at least, greatly impaired its momentum; and indeed, before the Transportation Service became completely organized, it did bring about a serious situation. The telegram, with its implication that the commander of combat troops still had much to do with the transport of his men, can also serve as a basis for the reader's appreciation of the tremendous changes brought about later in our handling of troops and supplies.

In the latter part of May the authorities were working with all speed to prepare the ships for this first sailing. The Government had found the vessels at or near New York and had bought or chartered them to be the beginning of what was later a mighty transoceanic equipment. The army ships of the first convoy, fourteen in number, were these:

Havana	*Montanan*
Antilles	*H. R. Mallory*
Dakotan	*Momus*
El Occidente	*Pastores*
Finland	*San Jacinto*
Lenape	*Saratoga*
Edward Luckenbach	*Tenadores*

From The War College Collection

BORDER TROOPS OFF FOR FRANCE

Photo by International Film Service

"GOOD-BYE, BROADWAY!"

From The War College Collection

AN EARLY TROOP TRAIN

Photo by Western Newspaper Union

SOME OF THE FIRST TO GO

With the single exception of the *Finland*, none of these vessels had ever been in the transatlantic trade. They were all boats from the coastwise and Latin-American trades, running from New York to the West Indies, to the American and Mexican Gulf ports, and to Central America. The *Pastores* and *Tenadores* had been in the banana trade between New York and Caribbean ports, as part of the "Great White Fleet" of the United Fruit Company. The *Mallory* was in coastwise trade between American Gulf ports, Cuba, and New York. Most of the ships were in New York harbor, either unloading or already unloaded, when the Government secured them. The *Luckenbach* was in Philadelphia unloading. She steamed around to New York, reaching there May 30, the date also of the arrival of the *Saratoga*. The *Momus* reached New York May 31, and the *Finland* came in from her last commercial voyage across the Atlantic on June 1.

While these vessels were receiving alterations to make them over into troop transports and cargo and animal transports, under the direction of construction officers of the Army Transport Service, the New York Navy Yard was preparing gun platforms and mounts for the decks and securing guns for the mounts, so as to give each vessel her own defense against the submarines which were expected to dispute the passage of the convoy across the ocean. The *Finland* alone was already armed, fore and aft.

On the day when war became a legal fact, the Government had seized the great piers of the North German Lloyd Steamship Company and the Hamburg-American Line at Hoboken, New Jersey, across the North River from Manhattan Island. Here the Army Transport Service set up headquarters and prepared for the embarkation of the First Division.

For the benefit of the morale of her civilian population, Great Britain invited us to send part of the First Expeditionary Division to England to visit London *en route* to France. Military and naval reasons forbade us to accept this offer of hospitality; but the Government promised that later contingents should land in England. This promise was fulfilled on a scale far greater, perhaps, than either our Government or the Government of Great Britain expected.

The four infantry regiments at the Mexican border and the other units were equipped and ready for departure on June 1, the date set in the original orders from Washington. The supply companies of these regiments, in fact, entrained at their various headquarters in Texas on the 31st day of May and reached the New Jersey suburbs of New York on June 7. By June 1 the refitting of transports had progressed to such a point that the War Department felt justified in ordering the Texas units to entrain at once.

The most remote of the four infantry regiments, the 18th, left its quarters at Douglas, Arizona, on June 2. The other three regiments, all of which were stationed in Texas, entrained on the 3d, each in six special trains. The 16th had been patrolling the border at the Rio Grande crossing at El Paso. It entrained between midnight and four o'clock on the morning of June 3. The 26th entrained at its camp at San Benito, Texas, in the late afternoon of June 3. The 28th entrained at its headquarters in McAllen, Texas, on the same afternoon. The first section of the train movement of the 28th started at 3.20 p.m., and the last left McAllen at 8.40 that evening, the others getting off at regular intervals between those times. The entraining of the other regiments was conducted on a similar schedule and with equal precision.

The 16th traveled *via* Fort Worth, St. Louis, and Buffalo, three sections arriving at Hoboken on June 9 and three on June 10, the entire regiment having made the trip from the Mexican border within seven days. And these troops carried all their equipment on the journey by rail—a fact which means that freight cars were coupled to the passenger trains, with a consequent slowing down of the running time. The 18th went *via* Chicago, reaching Hoboken on June 9 after a journey of seven days. The quickest trip of all was made by the 26th, which, routed through St. Louis, began arriving in Hoboken on June 7 after four days of travel. The last two sections carrying the 26th reached Hoboken on the morning of the 8th. The 28th Infantry arrived in Hoboken on June 8 and 9, *via* Atlanta and Washington, D. C. All four regiments were at the Port of Embarkation by June 10.

Meanwhile the other units assigned to the First Division had been traveling to Hoboken. Base Hospital No. 18 had been ordered to mobilize at Allentown, Pennsylvania, for departure with the first convoy. This hospital was the creation of Johns Hopkins University at Baltimore. Allentown is but a few hours' run from New York, so that the hospital unit could entrain there, travel to New York, and embark on a transport, all within one day. Two battalions of engineers assigned to the division left San Antonio in three sections on June 2, arriving at Hoboken June 9. The field hospital and ambulance company designated for the division occupied one train, which left San Antonio June 3 and arrived at Hoboken June 9. The Signal Corps outpost company's train left Brownsville on the 3d and arrived at Hoboken on the 9th. A detachment of quartermaster troops had also come through from Texas on a special train.

The embarkation of all these troops from the piers in the North River—piers recently German, by the way—was not so rapid as the execution of similar tasks later on. Approximately 12,000 troops sailed in the first convoy. They spent four days getting their baggage and themselves aboard ship. Hoboken was to see 45,000 American soldiers embark on transports within twenty-four hours.

The troopships of the convoy started June 14, 1917. The 16th Infantry occupied two transports—six companies and the regimental headquarters on the *Havana*, six companies and the supply company on the *Saratoga*. The 18th crossed on the *Finland* and the *Mallory*, six companies to each ship, with the headquarters on the *Finland* and the supply company on the *Mallory*. These assignments used up the larger ships, so that the 26th Infantry and the 28th Infantry each had to utilize three vessels. The 26th traveled on the *San Jacinto*, the *Lenape*, and the *Momus*, one battalion of four companies on each vessel. The headquarters of the 26th was set up on the

San Jacinto; the regiment's supply company found accommodations on the *Lenape*. The three ships assigned to the 28th were the *Antilles*, *Pastores*, and *Tenadores*. This regiment, also, embarked with its three battalions intact, one on board each of the three ships. Regimental headquarters traveled on the *Antilles* and the supply company was assigned to the *Pastores*. General Sibert, his staff, and the headquarters of the First Division were on the *Lenape*, sharing the vessel with troops of the 26th Infantry. The quartermaster troops were carried on the *Pastores*. The ambulance company and the field hospital embarked on the *San Jacinto*. The four motor truck companies, personnel only, had quarters on the *Finland*. The Signal Corps outpost company was on the *Mallory*, and 500 military stevedores were carried on the *Momus* and the *Tenadores*.

Thus, of the fourteen vessels in the first American convoy, ten carried troops—the *Antilles, Finland, Lenape, Mallory, San Jacinto, Saratoga, Momus, Pastores, Tenadores*, and *Havana*. The other four vessels, the *Dakotan, El Occidente, Edward Luckenbach*, and *Montanan*, carried cargo and animals for the First Division. This last group sailed June 17.

We must not forget the regiment of marines which crossed the ocean with the infantry of the First Division. The Navy, which protected the first convoy and had charge of it throughout the voyage, carried the marines aboard its own ships. The Navy already possessed the naval transports *Henderson* (new) and *Hancock*, and had also seized the German auxiliary cruiser *Printz Eitel Friedrich*, which had been interned in Philadelphia. This vessel was a passenger ship; and, since she had escaped the damage inflicted by the German crews upon the machinery of nearly all the German vessels sheltered in our ports, she was ready for immediate service as a naval transport. The Navy renamed this vessel the *De Kalb*. On these three ships, the *Henderson, Hancock*, and *De Kalb*, attached to the first convoy, the 1st Regiment of Marines started on the road to France.

Such was the distribution of the first combat troops of the

A. E. F. as they set out from America on that memorable June day—a gallant muster, destined to write its record imperishably into the history of the world. The orderly line of transports, some in sober gray, others bedizened like harlequins in the fantastic patterns of the new camouflage, pass through the opened gate in the recently placed submarine net across the Narrows; through, and on into the obscuring fog of a spring morning. There we wave our farewells to them for the present, reserving, however, a place later in the narrative for the thrilling story of their unforgettable voyage across the Atlantic.

Even before the first convoy sailed, the War Department was preparing for the departure of other overseas troop units which should make the First Expeditionary Division complete and place in France the skeleton of the Second Division. In the first convoy sailed only the infantry and a few miscellaneous units of the First Division. It was necessary to add to the First its artillery, its engineering force, and other necessary sections. No time was lost in preparing these organizations for foreign duty. As yet, our transport fleet consisted solely of the fourteen vessels of the first convoy. It was the plan to have the other units of the First Division at the docks in New York awaiting the return of the transports. At first the authorities thought that the ships could be ready to sail on the second voyage from America by July 15, and this date was tentatively fixed for the mobilization of more troops at the Port. But the arrival of so many vessels at one time seriously congested St. Nazaire, our first port of debarkation in France; and the 500 stevedores carried across on the first convoy were far too few to handle the work of unloading in quick time. July 15 came and went, and still the transports had not returned from France.

By July 4 the War Department had designated the troops which were to sail in the second convoy. The artillery for the First Division included the 5th Regiment, then stationed at Fort Bliss, Texas, the 6th, quartered at Douglas, Arizona, and the 7th, at Fort Sam Houston, Texas. A trench-mortar battery for the Division was being organized hastily at Fort Dupont,

Delaware. These three regiments of Field Artillery were to comprise a brigade, the headquarters of which was to be organized at once in readiness for sailing. The 1st Regiment of Engineers received orders to join the First Division in France. Its companies were located at various points in Texas, except one company, stationed partly in New York and partly in Washington, D. C. The regiment assembled in Washington and moved thence to Hoboken. The 2d Field Battalion of the Signal Corps, at Brownsville, Texas, which had sent its outpost company with the first convoy, was ordered to sail on the second convoy. A battalion of telegraph operators for the A. E. F. was recruited and organized at Monmouth Park, New Jersey. The Southern Department of the Army was ordered to organize the horse-drawn section of the ammunition train for the First Division, and the Eastern Department the motor-drawn section. Meanwhile the War Department was getting together a headquarters train with a company of military police for the First Division. The First Aëro Squadron and three base hospitals also found ship space, as did six railway engineer regiments. In all, the troops embarking on the second convoy numbered 274 officers and 7,337 men. They carried with them 797 vehicles of all sorts, including wagons, rolling kitchens, artillery caissons, motor trucks, automobiles, and motorcycles.

The second convoy crossed early in August. It sailed in two escorted groups of ships. The first group, which left on July 31, 1917, included four of the troop transports which had been members of the first convoy: the *Pastores*, *Tenadores*, *Mallory*, and *Saratoga*. The tanker *Arethusa* also sailed in this group, carrying fuel to the naval forces abroad. Six days later sailed the second group, composed of the *Finland*, *Antilles*, and *San Jacinto*, with the navy transport *Henderson*. This departure ended the embarkation at New York until September, when there began the unbroken flow of American soldiers to France which was to end only with the armistice.

CHAPTER III

MOBILIZING REGULARS AND NATIONAL GUARD

ALTHOUGH, from a popular standpoint, the chief interest in these early weeks of the war attaches to the inland travel and overseas embarkation of the First Division and other pioneer units of the American Expeditionary Forces, this phase of military movement constituted by no means the major part of the work then being conducted by the organization which was handling transportation. Military transportation on a war-time scale began almost as soon as war was declared.

For months the United States had been teeming with agents of the German Government, who operated with the weapons of the Vandal against American factories turning out munitions of war for the Allied armies and against American railroads transporting these supplies. There was no reason to believe that these hirelings would not continue their depredations after America became a belligerent, when the opportunity for outrages was vastly greater. There were hundreds of railroad bridges, the destruction of any one of which would seriously cripple railroad transportation at an hour when it was imperative that every mile of trackage be used to its capacity. For the protection of these structures and of important industrial plants, the Government turned out the National Guard. One of the first duties of the transportation service was to convey large numbers of state militiamen to the important bridges, tunnels, and industrial plants which needed protection. This movement, however, was not large enough to be considered at this point as one of the distinctive phases of military travel during the war. Of these greater drifts or tides that characterized different periods of army transportation, there are five which should have immediate mention:

(1) The movement of the old units of the Regular Army to increment camps;

(2) The movement of the National Guard organizations to their training camps;

(3) The movement of drafted troops from their homes to their cantonments;

(4) The intercamp movement of National Army troops; and

(5) The movement of all to the seaboard for embarkation. Only roughly speaking were these currents consecutive in point of time. They always overlapped. And for a long span the last three processes occurred simultaneously—that is, during the last ten months of military expansion new drafts were continually called to the cantonments, intercamp travel continued as these new levies were sorted and distributed, and in the same period the stream of troops flowing to the ports reached flood-tide and stayed there.

Yet one or another of these types of movement dominated the traffic at different periods. With reasonable accuracy we can define the limits of these periods as follows:

(1) Regulars to increment camps—late spring of 1917;

(2) National Guard to training camps—early autumn of 1917;

(3) Drafted men to cantonments—mid-autumn of 1917;

(4) Intercamp travel—late fall and winter of 1917-1918; and

(5) Movement to ports—March-October, 1918.

This classification, of course, does not include the tremendous volume of rail travel incident to demobilization after November 11, 1918. Nor does it embrace such miscellaneous, but heavy, items as the gathering of volunteer recruits from the enlistment centers; the carrying of candidates and officer graduates to and from the various officers' training camps; the transport of troops to depots, arsenals, dangerous war factories, army posts, hospitals, and other stations not on the

Photo copyright by International Film Service

A BRIDGE PATROL

Photo by International Film Service

CHICAGO GUARDSMEN OUT FOR WAR SERVICE

Photo copyright by International Film Service

NATIONAL GUARDSMEN OF NEW YORK LEAVING FOR CAMP

From The War College Collection

NEWARK (N. J.) FAREWELL TO NATIONAL GUARD

direct route between the citizen soldiers' homes and the A. E. F. in France; the tremendous furlough travel that weighted the rails during the Thanksgiving and Christmas holidays; or any of a dozen other elements that enormously added to the traffic burden and complicated its problem.

Finally, this classification does not include (nor is any of the discussion in these pages concerned with) the travel of individual soldiers or small groups of them. These isolated travelers arranged for their own trips and rode on the regular trains, the railroads handling them as commercial passengers. They presented transportation orders to the ticket offices and received their tickets, and the military disbursing office later paid the cost of their transportation. On furlough or leave, military passengers rode for one cent a mile and paid it from their own pockets. That this travel was heavy, anyone who rode on the trains in 1918 can testify. The transportation organization of the Army, however, assumed jurisdiction of the travel of soldiery only in groups of fifty or more, a party large enough to require at least one special car. The traffic figures presented in this account embrace nothing but officially managed travel, and do not include the casual travel of individuals.

The declaration of war was attended by a tremendous wave of patriotic enthusiasm, and men hurried by thousands to recruiting offices to join Uncle Sam's fighting forces. Whether from a desire to get at the Germans as soon as possible, or in order to escape the fancied stigma of being drafted for service, young Americans flocked to the recruiting offices in numbers never known before. They were taken into the military service in such numbers that it required special trains to haul them from the principal cities to various depots and barracks. In the five months after the declaration of war, and before the first selective service men moved to their cantonments, the transportation organization handled 125 special trains loaded with 33,277 of these recruits. This figure does not include the tens of thousands of volunteers from the smaller communities who traveled on regular trains to points of mobilization.

It devolved upon the transportation organization to move many of the old organizations of the Regular Army from stations along the Mexican border to camps in the East and South where they could receive these recruits and build up their own ranks to the strength authorized in the new scheme of army organization. This movement required many special trains; it was a difficult problem for the transportation organization, which was still relatively inexperienced. There were several of these Regular Army increment camps—one at Syracuse, New York, another at Westfield, Massachusetts, a third at Gettysburg, Pennsylvania, and a fourth at Chickamauga, Tennessee. In the West and Southwest the army forts served as increment camps.

Some instances of military travel in this period will indicate the nature of the movement. The 4th Infantry Regiment of the Regular Army left Brownsville, Texas, on five special trains on May 28 and arrived at Gettysburg on June 2. On May 28 the 7th Infantry entrained in four sections at El Paso, bound for Gettysburg, and arrived there June 4. On May 18 five sections started from San Antonio bearing the 9th Infantry to Syracuse, and reached their destination May 23. Chickamauga began receiving regular troops in late May, nine sections arriving between May 24 and 28, with the 6th Infantry from El Paso and the 11th from Douglas, Arizona. The 23d Infantry arrived in Syracuse on June 27 on four sections, having left El Paso June 19. The 30th Infantry left Eagle Pass, Texas, on May 19 on four special trains and arrived in Syracuse on May 25. Seven special trains, departing at intervals during three days in late May, bore the 11th Cavalry from Fort Bliss to Chickamauga. The 13th Cavalry moved in six sections from Fort Bliss to Fort Riley; the 3d Field Artillery traveled from points in Texas to Fort Myer in four sections; and the 4th Field Artillery went to Syracuse on three special trains—all three of these last-named movements taking place in May.

Between May 1 and August 1 the movements of the Regular Army in the United States, including the embarkation movement of the infantry units of the First Division, required the operation of 110 special trains, usually over long routes. The total number of troops carried on these trains was 36,765.

Among other troop movements in this period may be mentioned the dispatch of 10,243 officers and enlisted men, most of them volunteers for aviation, to various aviation fields in the United States and Canada, thirty-three special trains being operated for this one purpose in these early weeks. The recruiting, mobilization, and transferring of ambulance, hospital, sanitary, and medical companies before August required the operation of sixty-four special trains for 12,903 passengers. In this period the mobilization of engineer units required twenty-two special trains for 11,059 passengers. An interesting phase of early transportation history was the movement of casual officers, most of them graduates of the first officers' training camps. Thirty special trains, carrying 5,519 officer passengers, took part in this movement.

In all, the miscellaneous railway travel of troops, even before the transportation problem had become at all serious in its proportions, entailed the movement of 448 special trains with 138,482 passengers. This whole series of operations proceeded without the slightest disturbance to normal passenger train schedules. The public scarcely realized that troop travel had begun, so notably absent was the confusion which might have advertised the activity. Yet the number of men transported was comparable to the total force which received transportation at the time of the Spanish-American War.

Even so, the military passenger list of this period was negligible in comparison to the number of men that cascaded upon the transportation organization in August. In April, 1917, the National Guard had numbered approximately 150,000 men. By the middle of August, recruiting had added nearly 200,000 men to its rolls, so that its total strength was then above 340,000. The movement of this force to its training camps constituted the second great phase of inland passenger transportation during the war.

Sixteen training camps were provided for the National

Guard, nearly all of them in the South. This centralization of the camps in a single territorial section had an important bearing on the transportation problem, for it necessitated much longer hauls of troops than if the camps had been located in the centers of sixteen districts comprising the whole United States. The exigencies of the war program required that the Guard camps be set up in a mild climate. Winter was coming on. It was a certainty that not all of the National Guard divisions could be taken to France before the northern weather grew cold. Moreover, once the National Guard divisions had evacuated their training camps, there might be no other troops to take their places there. This dilemma was not to arrive in connection with the National Army divisions. The drafted men for the National Army were to be called out for training in continual increments, perhaps up to the whole sum of American manpower. Therefore, the National Army cantonments were of stanch, permanent construction, and were located according to the geographical distribution of population, regardless of climate. The National Guard camps were of but temporary construction—for the most part canvas tentage. To have placed these in the North would have put unnecessary hardship upon those who had to winter in them.

The southern location of the National Guard camps, then, placed upon the railroad system of the United States the heaviest military burden it had known up to that time. It required the transportation of great bodies of soldiers over great distances. Troops from Minnesota on the Canadian border went to Camp Cody at Deming, New Mexico, on the Mexican border. Militia traveled from Washington and Oregon to Camp Greene at Charlotte, North Carolina. The whole Guard movement called into being a complicated mesh of train schedules to bring the regiments together at mobilization points within their respective states and to carry them to their southern camps.

If military transportation were to break down at all, the disaster might have been expected at this time, when the National Guard, with its third of a million men, put the first of a series of military peak loads on American tracks. Yet the system never faltered. The hundreds upon hundreds of special trains, bearing their khaki-clad loads, went through to their destinations in almost as quick time as could have been made by regular trains carrying private travelers. The Guardsmen were still moving in great numbers when the first increment of the draft added its tens of thousands of men to the transportation load. These, too, were cared for by an organization now becoming exceedingly expert in its business.

American civilians were traveling in unprecedented numbers; the men in uniform were speeding along American rails in an ever-increasing flood; and those same rails were weighted to capacity with millions of tons of war freight—building materials for the training camps, raw materials for the thousands of industrial plants even then starting work on their war contracts, thousands of cars rolling toward the seaboard to bear to the Allies a volume of munitions which never for an instant dwindled by reason of America's entry into the war. At this juncture, in the fall of 1917, we see American transportation genius managing with supreme success a volume of traffic beyond the ability and equipment of any other nation.

The movement of the National Guard to its training camps was practically completed between August 15 and October 15, 1917. It involved the operation of 920 special trains loaded with 294,752 passengers. Approximately 50,000 National Guardsmen resided so near the training camps that they rode to them on regular trains or made their own way to camp without government transportation. An equivalent of this movement would be the travel of a single train over 710,309 miles of track, a mileage sufficient to take the single train nearly three times over the entire railroad system of the United States. Or, stated in another way, it was approximately the travel-equivalent of three troop trains operated over every short-line and trunk-line railroad in America. The Guard, in all, required rolling stock in such enormous quantities as 3,208 tourist cars, 3,941 parlor chair cars, 619 coaches, 1,211 baggage cars, 2,282 box cars, 1,097 flat cars, 478 gondolas, and

948 stock cars. The National Guard occupied on the first stage of the road to France a total of 13,802 cars.

New York placed aboard the trains a heavier human lading than any other state. New York's 37,787 National Guardsmen occupied 97 special trains. The Ohio National Guard was second in strength, with 24,065 men, occupying 76 special trains, bound for Dixie. Illinois sent 19,844 National Guardsmen to the training camps on 50 specials. The Pennsylvania Guard, 16,704 in strength, used 83 special trains. Wisconsin dispatched 47 specials; Missouri, 36.

In an appendix* to this volume is a tabular analysis of the primary travel performed by the National Guard, showing the number of men transported from each state, the number of special trains occupied, the period during which each state was sending its militia to camp, the destination of each movement, and the divisions of the American Army in which these units finally found place.

Although, in general, the National Guard troops traveled directly to the South to camps in which they were organized and trained in divisions, and from which they ultimately departed for the seaboard and the ships, there were exceptions to the rule. The National Guard of all the New England States, for example, assembled at Camp Devens, the new National Army cantonment at Ayer, Massachusetts, and there combined with the New England Coast Artillery to form the Twenty-sixth Division, its ranks being filled out by a slight addition of drafted men. The Twenty-sixth was known as the "New England Division." Almost immediately after its organization it began moving to the Port of Embarkation to cross the ocean and receive the greater part of its training abroad. By the 1st of November the entire division had landed in France, surrendering Camp Devens to the Seventy-sixth Division of the National Army, made up of drafted troops from New England and New York State.

The Twenty-sixth Division and the Forty-second (the famous "Rainbow Division") shared the honor of being the

* Appendix A.

first National Guard divisions to arrive as units in France. They immediately followed the First and Second Regular Army Divisions into the organization of the A. E. F. The "New England" Division was the second to go into the trenches and one of the first four or five to begin active combat operations against the enemy. The assembling of this organization in the heart of a small and heavily populated district put only a slight strain upon the transportation facilities.

The Forty-second, the "Rainbow Division," was the only other National Guard division which did not train in the South. It assembled and organized at Camps Upton and Mills on Long Island, near New York City. Camp Mills, twenty miles by rail from the ferries on the East River, was originally set up for the accommodation of the Forty-second Division, and was therefore, like the other National Guard camps, of temporary construction. Soon, however, it became evident that Camp Merritt, in the northern New Jersey suburbs of New York on the Hudson Palisades, and designated as the rest camp for France-bound troops awaiting transport space at the port of New York, would be inadequate, large and well equipped as it was, to handle the flood of men which must pass through the Port. The result was that, after the "Rainbow Division" had left Camp Mills for France and the evacuation of the camp was complete, Camp Mills was rebuilt with permanent barracks and added to Camp Merritt as part of the permanent facilities of the Port of Embarkation. Camp Mills began to be used as an embarkation camp early in 1918, and thereafter its capacity of 40,000 visiting troops was filled and refilled times without number.

The assembling of the Forty-second Division was a considerable task, for the division was made up of National Guard troops from twenty-seven states of the Union, with a considerable dilution of drafted men. Alabama furnished the largest contingent, and New York the next largest, followed closely by Iowa, Ohio, Indiana, Minnesota, and Illinois, in the order named. The concentration of these and the other units on Long Island, during the period of greatest travel by the National Guard, meant that every day there were a few National Guard special trains moving eastward across the southerly current of travel which was the dominant characteristic of military transportation in September, 1917.

Sentiment played a part in the composition of the Forty-first, known as the "Sunset Division"; and in that fact lies the explanation of one apparent inefficiency in planning the primary movement of the National Guard. As the training system was first formulated, it was proposed to assemble one National Guard division at Camp Greene, near Charlotte, North Carolina. The plan was to build this division of National Guardsmen out of the northwestern tier of states from North Dakota to Washington. The plan was carried to the point of dispatching a great number of special troop trains across the continent to Charlotte from points as far west as Puget Sound. Idaho sent four such trains, Montana two, North Dakota eight specials, Oregon twelve; Guardsmen of South Dakota occupied five special trains in this movement, those of Wyoming four; and six specials carried more than 2,000 Guardsmen from Washington on the Pacific coast to North Carolina on the Atlantic. Each of the trains from Washington and Oregon was on the rails for more than a week. Yet scarcely had these troops been set down in the new camp in North Carolina when there arose a popular demand in the Northwest for an entire National Guard division to be made up of units from the Northwestern States. The War Department deferred to this expression of public opinion by establishing at Camp Fremont, near San Francisco, the Forty-first Division, composed of National Guard troops. Among the states represented in this division were Oregon, Washington, Idaho, Montana, North Dakota, South Dakota, Wyoming, Colorado, and New Mexico. To assemble the division involved, of course, the return to the Pacific coast of troops that had already crossed the continent. And some of the northwestern troops originally sent to Camp Greene went to points other than Camp Fremont. Several organizations joined the "Rainbow Division" at Camp Mills. Certain of the North

Dakota and South Dakota units were transferred to Camp Cody, at Deming, New Mexico, there to become part of the Thirty-fourth Division, made up principally of Guard troops from the upper Mississippi valley. After the northwestern troops had evacuated Camp Greene, the establishment became an increment camp for the Third and Fourth Divisions of Regulars. The Third occupied the camp from late November to late March, the Fourth from mid-December to mid-May.

Two states, New Mexico and Nevada, had no National Guard organizations. Four states, New York, Pennsylvania, Ohio, and Illinois, each sent nearly enough National Guard troops to make up a division. The all-New York Division was called the Twenty-seventh. It trained at Camp Wadsworth, at Spartanburg, North Carolina. Every National Guard division received a greater or smaller infusion of National Army troops to bring it up to the authorized strength before sailing for France. The Pennsylvania Division, known as the Twenty-eighth, which trained at Camp Hancock, at Augusta, Georgia, came the nearest of any of the seventeen National Guard divisions to being composed entirely of National Guard troops; it had only a few National Army men in its organization. The Illinois Division, the Thirty-third, which trained at Camp Logan, near Houston, Texas, received nearly half its strength from the National Army. The Ohio National Guard contributed slightly more than three-quarters of the roster of the Thirty-seventh Division, the remainder being National Army troops. The Thirty-seventh Division trained at Camp Sheridan, near Montgomery, Alabama.

In the spring and summer of 1917, then, the National Guard, at its various camps and armories throughout the country, received volunteer increments, until about 200,000 new soldiers had been added to the ranks. The militia trained at the Mexican border numbered only about 110,000 men. This trained force was to be simply the skeletal nucleus of the great new volunteer army of Guardsmen. By the middle of August the southern training camps were ready to receive their first units, and the Guard began moving southward. Some states sent out trains as early as August 11; the heaviest movement occurred in September. During the middle two weeks of September, practically all the states were dispatching troops. The first trainload of National Guard troops from New York started for Camp Wadsworth on August 29. Thereafter the New York movement was rapid—three special trains on August 30, three on September 6, three on the 7th, one on the 8th, four on the 11th, four on the 14th, and five on the 15th. After a quiet interval of eight days, during which there was no National Guard travel from New York, eight special trains of New York Guardsmen departed for camp on September 24, eight more on the 25th, one on the 27th, five on the 29th, six on the 30th, and four on October 1. Another interval, and three more trains started October 6, one on October 8, eight on October 9, five on October 11, and three on October 12, which completed the movement.

Throughout this period there was a considerable employment of special trains for concentrating the various New York National Guard units at suitable entraining points within the state. When distances were short, heavy trains were the rule. The entire 69th New York Infantry Regiment, 1,716 men, rode from New York City to Camp Mills, Long Island, on a single train, made up of thirty-three chair cars, one baggage car, and eight kitchen cars—a total of forty-two.

The dispatch of Ohio National Guard Troops was another big job in transportation. The 4th Infantry Regiment of the Ohio National Guard, nearly 4,000 men, became part of the "Rainbow Division." Having gathered at Camp Perry, Ohio, this regiment started for Camp Mills on September 7 on eight special trains. The movement to the Ohio camp at Montgomery, Alabama, began on August 23. The 10th Infantry Regiment of the Ohio Guard, consisting of volunteers from northeastern Ohio, left Youngstown on three special trains September 16. The 5th Infantry left Cleveland and other northern Ohio points on six sections September 25 and 27. The 6th Ohio Infantry left Cleveland and Toledo on three special trains September 27. The 8th Infantry entrained on

Photo by Brown Bros.

FIFTH AVENUE CHEERS ITS DEPARTING GUARDSMEN

From The War College Collection

OFF ON THE LONG ROAD

Photo from American Red Cross

TRAVELING CANTEEN

Photo by Felix J. Koch

PUBLIC FAREWELL TO CINCINNATI NATIONAL GUARD REGIMENT

four specials September 29. The next day the 2d Infantry left northwestern Ohio cities on three sections. The 3d Infantry, which had concentrated at Camp Sherman at Chillicothe, embarked on five special trains October 8 and 9. Five other trains on October 9 bore the 1st Ohio Infantry from Cincinnati.

The movement of Pennsylvania National Guard troops to Camp Hancock, Augusta, Georgia, was so expeditious as to merit a detailed analysis. The Pennsylvania Guard began moving southward August 17, and by September 15 the entire division was in camp. Except for a few scattered units, the whole force of 17,000 men was transported from Pennsylvania to Georgia within eight days.

The heavy movement began on September 7, when nine special trains departed for the South, including four sections which carried the 10th Infantry Regiment from Pittsburg. The next day, September 8, the 16th Infantry left Meadville and Erie on six trains; the 18th Infantry started from Pittsburg on five sections; and the 3d Infantry entrained at West Pittston on three sections—fourteen National Guard trains, all running as specials, originating in Pennsylvania on that one day. Eleven of these trains discharged their passengers at Augusta on September 10, two days later, and the other three early in the morning of September 11. On September 10, twelve sections left Pennsylvania points for the South with various other units of the National Guard, including the 13th Infantry from Scranton, which traveled on five trains. The 1st Infantry left Philadelphia on September 11 in four sections. Seven special trains left the state on the 11th, six on the 12th, and four on the 13th, ending the movement. In seven days, September 7-13, fifty-two of the eighty-three sections used in the movement of Pennsylvania state troops to the training camp had been loaded and dispatched.

These trains fairly deluged their passengers upon Camp Hancock. On September 10 fifteen trains unloaded at the camp. The next day there were five more. On the 12th there were ten arrivals. September 14 was another big day at Camp

Hancock, fifteen special trains arriving and discharging their loads. The first arrivals occurred at dawn, the last after nightfall; and some of the specials pulled into camp at fifteen-minute intervals.

This whole movement of the National Guard to its camps—the transportation of more than a third of a million men—went through with the most gratifying expedition and smoothness. Largely because of this celerity in transportation, the National Guard divisions were ready when, in 1918, the desperate call for troops came. Every one of the seventeen National Guard divisions, at least in part, reached France. By the late spring of 1918, most of the National Guard camps in the South were tenantless.

CHAPTER IV

THE TROOP-MOVEMENT OFFICE

UP to the end of the year 1917, over half a million American troops had traveled in organized units upon American railroads. Their travel had involved the operation of nearly 1,500 special trains. These figures, large as they then seemed, and soon to be doubled in the inland military traffic of a single month, do not take into account the movement of several hundred thousand selective service troops inducted into the Army in the autumn of 1917.

The movement of a special train calls for a vastly greater amount of detail-work on the part of a railroad organization than is required for the operation of a regular train. The regular train has a fixed schedule, and it is handled as a matter of routine and familiar rule, both in the railroad office and on the tracks. The special is what its name implies—specially assembled, routed, and scheduled, with every employee concerned acting under special instructions. It is entirely outside the traffic routine, and the work involved is all added work. The operation of 1,500 such trains constitutes a prodigious transportation problem. No other country, no country with a railroad system of smaller extent and equipment than ours, or with railroad operatives of slighter skill and experience, could have handled this additional burden of traffic without disruption of the regular passenger and freight schedules. Not only "business as usual" was the rule on American railroads in 1917, but business much greater than usual. In the transportation of passengers the American railroads met every commercial demand put upon them by the exigencies of war, and in addition handled the great volume of purely military traffic. So far as the movement of troops was concerned, they did it rather easily, especially in 1917.

It was evident that we could be no stronger in France than our transportation system at home permitted us to be. President Wilson, in the most historic of his war papers, the message to Congress in which he declared that America was entering the war to make the world safe for democracy, called upon the men who run the railways of the country to see to it "that those arteries suffer no obstruction of any kind, no inefficiency or slackened power." A few weeks later, in Washington, the victor of the Marne, Marshal Joffre, turning to a group of American railway officials, exclaimed dramatically: "This is a railway war! The battle of the Marne was won by the railways in France!" By the end of 1917 it had become evident that American transportation was equal to the great task.

What had brought about the change? Why were the railroads of 1898, highly organized even then, as they were, unable to transport, except with great confusion, delay, and inefficiency, the mere handful of men mobilized in the war against Spain; whereas in 1917 those same roads, weighted with a regular traffic several times as heavy, could transport one of the largest armies ever organized over far greater distances, and that without friction or "obstruction" of the "arteries"? The secret lay in organization, or rather in a difference of organization. In 1898, the military and railroad organization in charge of troop travel was decentralized; in 1917, it was centralized and supreme. In 1898, military traffic was handled by district quartermasters who left it largely to the enterprise of the commanders of the troops to secure railroad facilities for themselves; in 1917, transportation was entirely in the hands of a central military traffic organization in Washington. In 1898, the railroads were split up into their independent units; competing lines bid for troop-movement contracts; the troops moved along the rail systems which had underbid their competitors; and the dispatchers of trains knew nothing of conditions at the terminals. In 1917, this method had been entirely abandoned; and, so far as the transportation of troops was concerned, all the railroads of the United States were operated as a single system administered through an organization formed by the railroads themselves in Washington—a volunteer organization with the power (and the ability) to direct every mile of troop travel from point of origin to destination, coördinating the entire troop transportation service as a master dispatcher might organize the passenger service of any one great railroad system.

To the troop-movement office of the American Railway Association and to the brilliant traffic executive at the head of it, the late Mr. George Hodges, belongs a large share of the credit for the unbroken and unimpeded movement of the great army in 1917 and 1918 to and between its training camps and to the Atlantic seaboard. Mr. Hodges was virtually unknown to the American public and even to the millions of troops who rode under his control, but no man was better known to the operating railroad men of the United States during the war. He was the field commander of inland military passenger transportation. The organization which he built up functioned with such precision that it was taken over bodily by the United States Railroad Administration when that body assumed operation of the American railroads in early 1918, and its functions and personnel were then left intact. Indeed, so admirable was the work of this civilian office that the Inland Traffic Service of the Army, instead of attempting to build up a complete troop-movement section of its own, contented itself with acting as the liaison agency, merely forwarding the commands of the executive head of the Army to the troop-movement office of the railroads, certain that those commands would be carried out on time and to the letter.

The time-honored system of letting troop-transportation contracts to the railroads which submitted the lowest bids was not abruptly changed to meet the situation of 1917. Even before the outbreak of the war in Europe the American War Department had taken first steps toward the consolidation of American railroad lines to meet a military emergency. As early as May, 1914, the Quartermaster General of the Army had written to the American Railway Association suggesting that the association maintain officers in Washington to consult with the military authorities on problems of transporting troops and supplies. After a conferenee the A. R. A. designated a representative to coöperate with the Army in all matters of transportation except routing, which was still to be left to the competition of the various railroads. At that time the disturbances in Mexico had created a situation which might at any time have culminated in war. Within a few weeks this original arrangement had assumed graver importance: war had broken out in Europe, and it was soon seen to be a war whieh might engulf the whole civilized world. From the beginning of the struggle in Europe, then, we had in the United States at least the nucleus of an organization eapable of welding all the railroads of America into a great strategic unit.

A year later the *Lusitania* had been sunk, and America had been drawn perilously near to the great eonflagration in Europe. On the suggestion of the Secretary of War, the railroads of the United States established in Washington, in August, 1915, a special committee on coöperation with the military authorities. Mr. Fairfax Harrison, the president of the Southern Railway, was the chairman of this committee, and the other members were Mr. R. H. Aishton, now the president of the Chicago & Northwestern Railroad, Mr. W. G. Besler, president of the Central Railroad of New Jersey, and Mr. A. W. Thompson, viee-president of the Baltimore & Ohio. The conferences with this committee lasted nearly a year, and a scheme of railway unification for military purposes had been well developed when, on June 18, 1916, the state militia of the United States was ordered to mobilize on the Mexican border.

Here was made the first use of the plans drawn up in Washington. In the border mobilization in 1916 the operating branch of the A. R. A. committee in Washington received the invaluable experience which enabled it, a year later, to function with such precision and ease. Mr. Hodges, at Washington, took executive charge of troop transportation. The dispatch of trains from various points in the United States to the Mexican boundary was entirely in his hands. He started trains for the Southwest according to the rate at which the terminals could receive them, thus preventing congestion at the ends of routes. He arranged facilities and connections so as to avoid delays, once a train had started. In all, 111,919 troops of the National Guard assembled at the border. The whole undertaking was handled so skillfully as to earn special commendation from the Adjutant General of the Army in his annual report for 1916.

We do not need to follow the intricate development of the American Railway Association's representation in Washington. As war with Germany drew nearer and finally became a fact, the original committee of four railroad executives expanded, undergoing several changes of name, until, in the spring of 1917, it possessed thirty-three members. It had an executive committee, unofficially known as the "Railroad War Board," composed of well-known railroad officials, and a smaller committee, which was known as the "general committee," though it might more accurately have been termed the "operating committee." This body consisted of Messrs. Fairfax Harrison, chairman; George Hodges, assistant chairman, and J. E. Fairbanks, secretary. Mr. Hodges was the member in charge of actual operations. The organization which he constructed was called "the troop-movement office of the Railway War Board."*

Meanwhile the Government had taken a long stride toward greater efficiency by abolishing the transportation contract. The National Guard and other troops had moved to the Mexican border under the old plan, some of the evils of which, however, had been largely offset by intelligently centralized control of the dispatching and arrival of trains. While the troops were still concentrated along the Mexican line, negotiations between the Quartermaster General of the Army and the representatives of the various passenger traffic committees resulted, on January 1, 1917, in an agreement whereby the

* It should be understood that this arrangement was concerned only with the transportation of troops. Military freight traffic was handled in quite another way.

railroads undertook to transport troops on order from the Government on a fixed tariff of fares. This agreement did away with the contract system. It enabled the Government—and hence the troop-movement office of the A. R. A., acting as the Government's agent—to use any railroad line or route in the United States the instant an emergency demanded such use, with no more waiting for bids to be submitted and contracts awarded. The military authorities could now prepare for actual travel as soon as any organization was ordered to move.

This agreement enormously facilitated the equitable common use of rolling stock. It virtually pooled all the transportation facilities and put them into the hands of the central operating organization, which was charged with the duty of dividing the traffic fairly among all railroad lines. That division, as a matter of fact, was bound to occur anyhow, for, in the event of heavy military traffic, it would become necessary to use all lines in order to accommodate the entire volume of traffic and avoid congestion on the main-traveled routes.

When war was declared, Mr. Hodges built up an organization to handle the technical details of the operating committee's job. In the Washington office he established three departments. One selected the routes for troop trains. Another, the transportation section, arranged for the actual movement of trains over the railroads involved, fixing the dates for the departure and passage of special trains and keeping in touch with all the necessary physical elements of transportation. This section also kept the record of troop movements. The third department was known as the Pullman section. Established originally by the Pullman Company as part of the troop-movement office, it had direction of the entire American equipment of tourist cars and apportioned them as they were needed. Later, after the whole organization had become a branch of the United States Railroad Administration, the Pullman section assumed control of all military passenger train equipment in the United States—coaches, baggage cars and

Photo by International Film Service

INSIDE A TROOP COACH

Photo by Paul Thompson

"GOOD-BYE, BOYS; GET THE KAISER!"

Photo by C. J. Fennel

A NEBRASKA TOWN SAYS FAREWELL

From The War College Collection

NEW ENGLAND CROWDS WATCH THE TROOP TRAINS PASS

express cars, kitchen cars, and also parlor chair cars and tourist sleepers.

This simple organization maintained contact with the War Department through the Department's Inland Traffic Service. The direction of troop movements was, of course, a military matter. The orders came from the Army. The details were left to Mr. Hodges and his troop-movement office.

Besides its headquarters in Washington, the troop-movement office maintained a field service covering the United States. It established a general transportation agent at each of the six department headquarters of the Army. Next, agents were placed at the increment camps of the Regular Army, after that at the National Army cantonments and National Guard camps, and finally at all embarkation camps, ports, and special camps at which considerable numbers of men were entrained and detrained. These agents carried out the orders of the central organization in Washington. It was the specific function of the camp agents to see to it that sufficient railroad equipment arrived in time for prospective departures, that the equipment was in good order, that the troops were properly loaded aboard the trains, and that the specials were dispatched promptly. These camp agents worked out a scientific system of loading troops, and they supervised the entraining whenever it occurred, day or night. The general departmental agents kept in touch with the various railroads, took charge of trains at junctions, and made sure that they were delivered expeditiously from one railroad system to another and that the arrangements were adequate for the prompt passage of troop trains across their particular districts. As it has been expressed, the duties of the field service were "to translate into terms of action the orders and necessities of the Army."

The general agents at the camps and other points were 127 in number, each with an office and the necessary clerical force. Each railroad in the United States designated one of its general officers, usually a vice-president, to direct the movement of troop trains on its lines and to coöperate with the railroad traffic agent at the army department headquarters. These officers—there were 204 of them—thus became virtually members of the troop-movement office, and helped further consolidate the American railroads into an operating unit.

With this organization the most extraordinary feats of military transportation were carried out with precision and certainty. In all, the troop-movement office handled upwards of 15,000,000 soldiers on special trains.

The man who built up this organization had spent his life almost from boyhood in railroading. He was the son of a Baltimore clergyman. In 1886, after an education received in New Hampshire, he entered the service of the Erie Railroad. In his career of thirty-one years as a railroad man he held responsible positions with such roads as the Baltimore & Ohio and the Seaboard Air Line. In 1910 he became a general representative of the railroads in various capacities connected with traffic and its relation to legislation, and in other public contacts. He died suddenly in Washington on March 14, 1919. By direction of the President, the Distinguished Service Medal was posthumously awarded to Mr. Hodges "for specially meritorious and conspicuous service as manager of the troop-movement section of the division of operation of the United States Railroad Administration. Mr. Hodges arranged all the details of the movement of troops from local draft boards to mobilization camps, between camps, or from mobilization camps to the ports of embarkation for shipment overseas. Troops in large numbers were moved on short notice, and he was responsible for the successful coördination and carrying out of these movements."*

In the chain of organization leading back to the source of authority, next behind the troop-movement section (which was the railroads' own organization) lay the Inland Traffic Service of the Division of Purchase, Storage, and Traffic. The Division of Purchase, Storage, and Traffic was a branch of the General Staff of the Army. The formation of P. S. & T.

* After the death of Mr. Hodges, his place was taken by Mr. C. F. Stewart, who successfully directed the tremendous volume of troop-passenger traffic incident to the demobilization of the Army in this country and the return of the A. E. F. from France.

was made imperative by the immensity of the war in which America was engaged. The division not only embraced the functions of the former Quartermaster Department, but it also appropriated almost every purchasing activity of the entire Army. It bought for the military service nearly everything from shoe strings to locomotives. It stored and distributed these supplies, and, as its name implies, it also transported them. It transported all troops as well. The Inland Traffic Service was the P. S. & T. Division's branch and agency for conducting the movement of troops and supplies upon the rails of the United States. Another branch of the P. S. & T. Division was called the Embarkation Service. It handled the troops at the ports, and also directed the operation of the army transports. Eventually both these organizations, the Inland Traffic Service and the Embarkation Service, merged into a single unit known as the Transportation Service, in which all the traffic functions of the Division of Purchase, Storage, and Traffic came together under a single executive.

The Inland Traffic Service maintained an extensive organization for handling the army freight traffic. The railroads themselves failing to unify military freight transportation, the Army took this problem into its own hands. The army troop-travel section, however, was simple, thanks to the extraordinary efficiency of the railroads' own organization.

Behind the troop section of the Inland Traffic Service, along the route through which authority over transportation must pass, lay still another organization: the Operations Division, also of the General Staff. The Operations Division was the direct agency of the Chief of Staff, who, in respect to the movement of men and supplies, was the contact point between the Army and its civilian control. When, out of the conferences of the President and his advisers, emerged policies, these policies, in so far as they related to the assembling and transportation of the Army, were transmitted by the Chief of Staff to his Operations Division, which body of experts translated the policies into general terms of action.

The Operations Division built the military programs. It

determined the number of divisions to be organized and the number of troops to be recruited, organized, and trained. It decided what numbers of men were to be called into the military service through the draft and what classes of men were to be called at each induction. It directed when the men were to be drafted, where they were to be sent, where assigned afterward. It called into being and organized all new units. It determined which units should be shipped overseas; it established the priorities in troop shipments. This office was one of the most vital branches of the war organization; for to it the Provost Marshal General, who had charge of the drafts, turned for the authority on which he based his plans. So, too, all of the supply bureaus of the War Department were dependent upon the Operations Division for information without which they could not prepare intelligently for the future. Almost every act of this division resulted ultimately in military transportation. With the transportation organization the Operations Division maintained intimate relations.

The sum of these facts reveals to us in its entirety the process of building and transporting a mighty army. That process is like the passage of an ingot through a steel mill. From the alembic of the high councils comes the molten metal—policies, fluid as yet, indeterminate outlines, round numbers. These go to the Operations Division and emerge hardened and welded into programs with definiteness in numbers and balance in proportions—so many combatant troops to be formed and trained, so many for replacement, so many for the Services of Supply. The assembling and the ultimate movement of this force toward France mean transportation. Orders go to the Inland Traffic Service; dates and places are fixed. We have progressed far from the general toward the concrete and specific. Finally, authority is relayed to the troop-movement office, and at last the grand strategy of the Government has become action, equipment, schedules, connections, rails vibrant with ten thousand wheels, thundering trains, windows crowded with cheering soldiers—the fruition of the nation's military plans.

CHAPTER V

HAULING THE SELECTIVES

IT has been necessary to tell something about the organization in charge of American troop travel in the war, for the reason that we come now to a subject which cannot be made clear until the reader understands the system which moved our soldiers in inland transportation. The men called to the colors by the Selective Service Law occupied, so far as the transportation organization was concerned, a different status from that of either the National Guard or the Regular Army. National Guardsmen and Regulars were part of the military organization before they boarded the trains, and it was the duty of the troop-movement system to conduct every phase of their progress, from entraining to destination. On the other hand, the great democratic army raised by the Selective Service Law was the gift of a free people to its Government and its military organization; and therefore the people themselves assembled the young selectives and delivered them to the Army at certain designated places—to begin with, the sixteen National Army cantonments. Theoretically, the military transportation organization had no part in the mobilization of the National Army at its cantonments. It handled National Army troops on the rails only after they had donned the uniform and taken their places in the organized ranks.

But only theoretically. In actual practice, the Army engineered the transfer of the hosts of civilian soldiers from homes to cantonments. This service took the form of volunteer assistance. The travel of conscripted men to camp rested ostensibly in the hands of the civilian state authorities; but the prompter behind the scenes was the Army itself, so ordering and organizing the whole enterprise that this vast tide of military travel should flow as smoothly as any other main current of the tremendous human flux.

The Selective Service Law placed on the administration of each state the duty of registering, exempting, calling out for military service, and transporting to the military centers the men within the scope of its provisions. But to have permitted each state to devise and carry out its own drafting system would have been to risk confusion and the failure of the law to do what Congress intended. Consequently the law provided that the Provost Marshal General should maintain supervisory control over the entire operation of the draft. He enlisted as his aides the adjutant generals of the various states, consulted with the governors as to the appointment of local and district boards of registration and exemption, set the dates for registering and induction, provided blank forms for the whole enterprise, and in general gave to the draft system that scientific coördination without which it could scarcely have been a success. And, also in an advisory capacity, the Provost Marshal General took charge of the transportation of the inducted men to their cantonments. It is for this reason that, in the cycloramic spectacle which the mighty citizens' army constituted on the initial stage of its journey to France, the Army's own travel bureau, the Inland Traffic Service of the Division of Purchase, Storage, and Traffic, does not appear. But we do find in a leading rôle that most important adjunct of the Inland Traffic Service, the troop-movement office of the American Railway Association. This civilian organization, acting directly in coöperation with the Provost Marshal General, arranged to the minutest detail for the mobilization by rail of the National Army.

Never had there been such an excursion before, never such a challenge to organized transportation. The assembling of great national conventions, or even the movement of such military forces as we had known up to that date in our history, were elementary traffic problems in comparison. Here was a task that involved, not a few main trunk lines, but the entire railroad network of the United States. This was special transportation, not from a few score or even a few hundred stations, but from thousands of them; from wherever, in short, there was even a single man waiting to be taken to a cantonment. The troop-movement office accepted complete responsibility in this undertaking, save for one small particular. It asked merely that the draft boards assemble their men at county seats or other local centers that possessed railroad facilities. Thus, almost without exception, it picked up the millions of selectives within walking or driving distance of their own homes.

To accomplish such a result, the troop-movement office built up a special organization, one which functioned almost automatically. The Washington headquarters appointed an officer to serve in collaboration with the governor or adjutant general of each state, and to act as the passenger traffic expert in charge of the travel of drafted men from that state. This officer was selected not only for his wide practical experience and ability in passenger transportation, but also for his temperamental adaptability to a position in which he must serve both the railroad-military and the civilian authorities. The ideal representative was a man of breadth of mind and vision, unprejudiced by his past associations in favor of any one railroad line or system, and able to view the entire trackage of his executive domain as a single unit. The presence of such a man in the capital made the state administration to which he was assigned expert in the manipulation of traffic. With such skill in the forty-nine headquarters, there was little likelihood of the delays and misunderstandings which were certainly to have been anticipated had the undertaking been attempted by novices, working out traffic problems in consultation with the railroads.

The minutiæ of operation had to extend to every railroad station at which even a single inducted man was to be entrained. There were over 4,500 such points in the United States. At each draft call, the troop-movement office prepared in advance a printed schedule for the entrainment and transportation of every man included. In compiling these schedules, the office called upon the railroad passenger traffic associations of the United States. The system worked approximately as follows:

Through its direct channel to the Provost Marshal General in Washington, the troop-movement office received its broad general instructions for any impending movement of selectives—the date on which the movement must begin, the number of men called out, the period within which the transportation must be completed. Turning to its traffic representatives at the state capitals, the troop-movement office obtained the details—the number of men to be entrained at each station in that state, and the destination to which each selective was to proceed. This mass of information was then divided into groups geographically and placed for action in the hands of the passenger traffic associations. Since each association was composed of the general passenger agents of all the railways in the association's territory, the members of each had at their fingers' ends the necessary data for a schedule covering that district. Wherever it was possible, the inducted men were sent aboard regular trains, usually in special cars. But when the times of departure of regular trains were not convenient, and whenever the confluent traffic had swelled to abnormal volume, the movement was conducted upon special trains. The traffic associations appraised their operating facilities and built up their draft schedules with regular or special train movements, as their particular exigencies prescribed. It was thus, by a system of both regular and special trains bringing the selectives together, first in small trickles like headwater rivulets, the volume constantly growing until finally broad currents were debouching into the mobilization camps, that the entire draft movement was handled. Everything was scheduled through in advance, with no detail overlooked. The central office in Washington arranged for the passage of troop trains across the boundaries of the traffic territories.

It is evident that there was a tremendous amount of detail involved in the preparation of a draft-travel schedule. The task amounted to the creation of an emergency passenger train service for the entire United States, which service was to function for a few days and then disappear. It might naturally enough be supposed that preparation for one of the great entrainments would require months. But so proficient did the troop-movement office become in its work that, on the fourteenth day after its receipt of notice of an impending general call for drafted men, it could be ready to transport these men by hundreds of thousands. In that fortnight it collected every needful detail of information, prepared schedules showing the exact minute of the day when each man was to board his train, printed the schedules, and distributed them throughout the United States, placing a copy in the hands of everyone concerned, down to the local board member and the leader in charge of each contingent of inducted men. Not only that, but ordinarily it completed the printing and distribution of schedules five or six days ahead of the first departures, so as to give the travelers-to-be plenty of time to familiarize themselves with the arrangements. The secret of this speed was, in a phrase, central control of decentralized organization.

Near the end of this record may be found a reproduction* of the schedule for the entrainment and movement of 3,500 drafted men of Minnesota to Camp Forrest, a special training camp at Lytle, Georgia. The schedule shows implicitly the solicitous oversight given by the central transportation authorities to the induction of the National Army into its camps. This particular movement was one of the last to occur. Military transportation was then at its highest efficiency. These 3,500 selectives were collected throughout Minnesota—even from as far north as the Canadian border—and carried to Georgia and set down in camp within a period of five days. The schedule for this movement was so plain and so lucid that any man in the contingent who could read at all must infallibly have understood it. The average railroad time-table issued for public consumption is a Chinese puzzle by comparison. Each Minnesota station was given its separate place in

* Appendix B.

the schedule, with the time of departure of the train and its time of arrival at the first important center. The schedule showed the county seat or railroad town to which the drafted men of each county were to go to entrain. It showed the number of men expected to entrain at each point, the railroad route to be followed through to the camp in the South, and the detailed time-table from the particular individual's community to the first principal concentration point. It even indicated what arrangements had been made for meals *en route*.

The feeding of selectives on the way to camp was a sizeable job in itself. It was the province of the Army to feed all other traveling troops, but the individual carriers supplied meals for the selectives. When inducted men were to ride for only short distances, the railroad usually provided box lunches for them. The Minnesota schedule shows that these men ate at station restaurants, at hotels, and on dining cars attached to the trains. This was a typical arrangement. The dining car was the rule on the special train carrying inducted men, but the railroads might provide as they found expedient. When a railroad possessed a good chain of station restaurants and could arrange to have its trains arrive at those restaurants at suitable hours, it sometimes fed its draft passengers there. The government authorities allowed two-hour intervals, at morning, noon, and night, for meals, and insisted that the soldiers be fed within those periods. The contingents of drafted men traveled under the direction of leaders, each appointed by the local draft board. The leader carried for each member of his party a railroad ticket and a sufficient quantity of meal tickets, which he issued as needed. Each meal ticket was worth sixty cents; it was accepted as legal tender on railroad dining cars, in railroad restaurants, and wherever the railroads had arranged for troop feeding; and local quartermasters redeemed it later at face value.

The first call for drafted men summoned 687,000 of them from 4,531 railroad points in the United States. At the time the call went forth the sixteen cantonments were far from ready to receive their future tenants. At some of them the

From The War College Collection

RAILROAD BOX LUNCHES FOR TRAVELING SELECTIVES

From The War College Collection

DRAFT TRAIN LEAVING A RHODE ISLAND CITY

From The War College Collection

A NEW ENGLAND TOWN DINES ITS INDUCTIVES

Photo by Kansas City Post

WHEN THE WAR BEGAN TO STRIKE HOME

work of construction had not been in progress for more than four weeks; yet the military program required such a stage of completion by September 1 that each of the sixteen centers should be ready to house one-third of its 40,000 future tenants in barracks of stanch construction, and also to give the men such sanitary necessities and public utility conveniences as are enjoyed only by the inhabitants of cities. It is to the credit of the Construction Division of the Army that the work was almost completed in the short time allowed. The movement of drafted men to the cantonments was not greatly delayed by the non-completion of the quarters, except at one or two camps where labor shortages and railroad congestion had hindered the work.

On August 13 the Provost Marshal General specified that, of the first draft of 687,000 men, 200,000, or 30 per cent, should entrain in a period beginning September 1, another 30 per cent beginning September 15, and a third 30 per cent September 30, the rest to follow as soon as was practicable. This plan was later so modified as to bring out the first contingent on September 5. This was done in order to avoid the Saturday half-holiday, Sunday, and Labor Day, three of the first five days in September. Soon it transpired that it would be unwise to burden the transportation lines with this heavy load of drafted troops during the very height of the travel of the National Guard to the southern training camps; and on August 25 the call for drafted men was again modified so as to bring out 5 per cent beginning September 5, and thereafter 40 per cent beginning September 19 and 40 per cent beginning October 3, the final 15 per cent to start for the cantonments on October 17. With some slight exceptions, this plan was carried out.

The first men of the National Army left their homes on September 5, and the movement was completed on September 9, exactly as scheduled. Only 5 per cent of the first draft, 35,000 men, moved in these five days, at the rate of 7,000 a day. Because of the slightness of the movement, divided as it was among fourteen destinations (neither Camp Upton, at

New York City, nor Camp Meade, near Washington, was yet ready to receive troops), it was accommodated entirely upon the regular passenger trains. Not until September 19 did the drafted men begin flowing into the cantonments in great numbers. Then were depicted scenes which will forever live in the memories of those who witnessed them.

The predictions of those opponents of the selective service system who prophesied that the conscripts would be dragged from their homes into the military service like shame-faced culprits, were shown to be utterly false. The eager spirits who had volunteered went to war amid no such scenes of celebration and patriotic emotion as heartened the men inducted for service in the National Army. In the cheering throngs which crowded the railroad stations as the trains bore off the young men of the land, there was doubtless many a heartburning at the thought of relatives who had volunteered and gone away officially unhonored. The departure of the selectives from almost every community in the United States was made the occasion for formality and ceremony. Towns were in holiday attire for the occasion, buildings draped with the national colors, stores, offices, and factories closed; and people thronged the streets. The inducted men gathered at the quarters of their local boards. Photographers were on the spot to snap the official pictures of the selectives, who lined up in as soldierly a posture as they know how to strike. The chairmen of the boards made speeches full of patriotic ardor, reminding the men of the great and solemn duty ahead of them, and incidentally calling their attention to the regulations prescribed for their journey to the training camps. There were other addresses, perhaps, by the great men of the town; and then the selectives went to the railroad station, either riding in gaily decorated automobiles or marching on foot, preceded by the best band the community afforded, and sometimes escorted by the veterans of other wars. Passing through the streets, the men heard such cheers as greet only heroes. There was nothing in the induction that savored of duress or the compulsion of the unwilling. These men started on the road to France as high-heartedly as the most fervidly impulsive of volunteers; and, indeed, there was among them many a man of the volunteer spirit who had preferred to wait for this hour and take comradeship in the most democratic army the nation had ever sent forth. Through cheers and tears, laughter and weeping, the din of horns and shouts of encouragement or bantering derision—for young America laughs most easily in moments of solemnity—the selectives made their way to the railroad station, followed thither by half the town, which would remain to cheer and bid Godspeed to the inducted men until the train had gone on.

Unfortunately, local exuberance and pride did not always take such innocuous forms. There were some who fancied that parting could not be duly celebrated without the aid of strong drink. In one western city from which the departure of the local contingent occurred on a Sunday evening, the over-enthusiastic mayor allowed the saloons to remain wide open. The saloon-keepers entered into the spirit of the day and gave the selectives *carte blanche* among their shelves. Many of the men of this contingent boarded the special train, not only with all the intoxicating liquor they could hold inside them, but with all they could carry in their arms besides. There were other instances of this sort, and distressing results followed—property destruction and the general wanton vandalism of intoxication—until finally the Government itself was forced to take cognizance and act.

The story of this phase of the transportation of drafted troops is incomplete without a reference to the classic journey made by a trainload of Arizona selectives in the days when drunkenness and departure for a national camp sometimes went hand in hand. The train, a special, carried a wild and untamed motley of passengers—cowboys from the great ranches, Mexicans, sheep herders, prospectors, desert dwellers, Indians, hard rock men from the copper mines, adventurers, business men, school teachers, mining engineers, and men from the East who had originally sought the Southwest for their health and been mended by the salubrious climate of the picturesque country from which the Selective Service Law took them. Aboard the train were a lawyer and several graduates of the eastern universities. And from millionaire to horse wrangler, the selectives aboard that train had one trait in common: they were drunk, and not just drunk, but extravagantly and supremely drunk. Few men aboard had escaped the contagion. The exhilaration and half-terror of going at last into the greatest of adventures, joined with the superinduced exaltation of alcohol, had put these blithe spirits "on the top of the wave."

There was in their travel little of the bestiality that branded some of the other drunken trips of the period. The journey of the Arizona contingent was an escapade in rough playfulness, conducted with a gravity in keeping with the occasion. It is true that once the men threw an offending porter overboard in the middle of the night and that some conscientious soul hurled after him a blanket in which (in the slightly improbable event of his surviving the fall) he could wrap his unclothed form against the chill of the mountain air; it is true that a fury of gambling blocked the aisles, and that there were sporadic fights and a continuity of profane speech. But these were the extremes.

The selectives had boarded the train in Arizona considerably, as an observer reported it, "the worse for wear." Admiring friends, men of their own boisterous stamp, had plied them with stirrup cups ere they departed to meet the Hun, and few had been able to resist the hospitality. The draft authorities at the various entraining points, however, had taken the precaution to confiscate all the inductees' bottles before they got on the train. The next morning, when the passengers awoke, they found themselves high and dry in Colorado—more particularly, dry. A great thirst pervaded the train from baggage car to tail lights. The morning was at its coldest and grayest to the selectives when, as if in response to their mute appeal, the train stopped at a Colorado station; and there across the dusty street, with a knot of cow ponies tied to the rail in front of it, was a screened emporium proclaiming on its signs that therein were buyable malt, vinous, and spirituous liquors. With glad shouts the passengers evacuated the train and dashed to the common goal. A few minutes later, when the distracted proprietor saw the last of his unwelcome patrons depart through the swinging door, he was also witness to the disappearance of the last bottle of his stock. The place had been looted clean.

Thereafter things livened up on the train. One favorite diversion of the selectives was to climb up on the roofs of the cars whenever the train stopped. The trainmen protested to them that, because of tunnels and overhead bridges further on, the train could not be started when loaded in such fashion, only to be met with counter-argument and rebuttal. At length, however, reason prevailed, and the train took up its riotous course. Lest their arms lose cunning for want of exercise, the cowmen got out their lariats at Trinidad, Colorado, and began to practice, with brilliant success, upon innocent bystanders on the station platform.

Certain wild fellows who had been living lonely lives in the interior regions of Arizona had not been able to bear the parting from their pets, and they had solved the problem by bringing the pets with them. One prospector had fetched along a tame wildcat—tame, that is to say, to its master, though it spat and glared ferociously at all others. Another car possessed a scarred bulldog; a third was proud in the ownership of an oft-embattled goat. As time went on, considerable contention arose over the respective fighting qualities of these animals, and eventually it was decided to settle the question of supremacy in a contest, a battle royal, in which the wildcat, the goat, and the bulldog should engage simultaneously. This affair was staged on the station platform at La Junta, Colorado. History does not relate on whom the honors of battle rested. Perhaps the spectators, with the fickleness of drunken men, forgot the engagement and turned to other affairs. The goat, at least, could not have been worsted, for at a later stop his admirers descended from their car and purchased five gallons of ice cream for the animal's supper.

There were other pets aboard which laid no claim to eminent pugnacity. One car flaunted a jackrabbit, another some puppies, a third three chickens which spent most of the time roosting dejectedly upon the coat hooks. This carload of celebrants insisted upon special privileges, and at intervals kept the train waiting while the chickens were exercised. The exercise for the fowls was largely vicarious: it consisted of riding on a broom handle carried up and down the station platform by one of their solicitous well-wishers.

One car alone possessed no mascot, and the inhabitants thereof moodily determined to rectify the disparity. When the train stopped at a small Colorado town, the passengers descried near the station an automobile of the best known of all makes. Commandeering this machine, a few of them drove down the main street with all the speed they could coax from the engine, a cowboy with swinging lasso on each running board. Presently the driver saw ahead a white bulldog sitting at the edge of the sidewalk. He ran close, and the cowboy on the near running board roped the dog and dragged him howling into the speeding car. The dog chanced to belong to the local constable, who emerged from a building with haste, cranked up his own Ford, and gave hot pursuit. He caught the marauders at the station, but it required his full powers of persuasion before the selectives would surrender the dog.

The Arizona incident and others of a more destructive import brought about a change in the method of handling selectives to cantonments. As the law was first administered, the drafted men did not officially become members of the Army until they had reached camp and got into their uniforms. In the Northwest several trains made their way through to Camp Lewis in the state of Washington to the accompaniment of most disgraceful and even bloody scenes. Railroad equipment suffered severely on these occasions. The Government determined to stop the trouble, of which intoxicating liquor was nearly always the proximate cause. Congress had already passed a law making it an offense to give or sell liquor to a

Photo from American Red Cross

THE MARCH TO THE RAILROAD STATION

From The War College Collection

A MONTANA STATION CROWD WHEN THE DRAFT TRAIN LEFT

Photo copyright by International Film Service

PARADE OF FIRST NEW YORK CITY DRAFTED MEN

Photo by Paul Thompson

ARM BANDS IN LIEU OF UNIFORMS FOR SELECTIVES

man in the military or naval establishments. The Provost Marshal General changed the mobilization regulations to provide for arm-bands, or brassards, stitched on the sleeves of selected men at the time of their induction by the local draft boards. This practice put the men technically in uniform before they said farewell to their fellow townsmen, and it effectively ended the disorder aboard the trains.

The Y. M. C. A. placed its welfare workers upon most of the draft trains. These workers, usually well selected, were able to do helpful work. At times, however, unwise men attempted to supply spiritual ministrations when the occasions were not auspicious. Such a one was the war worker aboard the Arizona train. To his report of what had occurred on this trip he appended the indignant comment: "Do you think we could get a song service or religious meeting with this gang? No, sir! All we could do was personal effort." The war workers accompanying selectives to camp rendered written accounts of each movement to the Y. M. C. A. War Board. The more picturesque aspects of the excursions may be reviewed in these files, which contain many an amusing, dramatic, or pathetic anecdote.

On a train moving westward from Danville in the Blue Mountains of Virginia, the Y. M. C. A. man was distressed by the wailing and sobbing of a young mountaineer, a veritable giant in physical proportions. Sobs shook his great bulk of bone and muscle, two hundred pounds of it, and his blubbering could be heard from one end of the car to the other. The welfare man, attempting to comfort him, learned to his surprise that the youthful giant was no craven: his outburst was not due to his fear of going to war. He had arranged to meet his sweetheart in Danville and to marry her before departing for camp. In some manner she had lost her way and failed to meet him, and to the homesick youth it seemed that he was bidding her good-bye forever. The Y. M. C. A. man comforted him with the suggestion that later on he could return on a furlough and see his inamorata after all. The young mountaineer's grief was stilled, and presently, as the train moved ever westward and never reached the jumping-off place, he, who had never before in his life been ten miles away from home, grew pop-eyed with admiration of the extent of the land. It was about fifty miles west of Danville that the big voice boomed forth to all who would listen: "Bud, if this old world is as big the other way as she is this, she's a hell-buster for sartain."

The first draft was handled by American rail transportation practically as scheduled. Because not all the cantonments reached 100-per-cent completion at the same time, there were some modifications in the original call. Also, after the drafted men had started, the War Department adopted the policy of bringing white and colored selectives into the cantonments at different intervals, and this action brought about some changes in the original plan. Moreover, not all the selectives included in the first call were sent to the cantonments: some were diverted to the coast defenses of the country. But by the end of 1917 all of the 687,000 men of the first draft were in uniform.

Thereafter the transportation of drafted troops from their homes to their training camps increased in volume month by month. The increase was especially marked in the spring of 1918, when the German Army began its supreme campaign for Paris and the English Channel. Throughout this period the regular movement of troops was also expanding at a tremendous rate. The movement of selectives was always conducted as an operation entirely apart from other troop transportation, but so efficient was the control of all operations that at no time was there interference. The equipment of rolling stock and tracks, pooled in a single system after January, 1918, was so manipulated that all military transportation went through promptly and in its proper order, and at all times the civilian public found a railroad system at its own disposal. The troop-movement office was called upon to handle as many as 50,000 drafted men in a single day, and a month's movement of selectives ran as high as 400,000 men.

The transportation of drafted men from their boards to the camps constituted nearly a quarter of the total military travel during the offensive period of the war. The draft passengers numbered over 2,750,000. These men rode an average distance of 388 miles to reach a mobilization camp. The composite haul was the equivalent of transporting the population of Chicago to Minneapolis. Reduced to terms of travel by a single passenger, the distance covered was well over a billion miles. The total military passenger mileage up to November 1, 1918, was nearly four and a half billion miles.

The policy of the War Department throughout the great mobilization was always to move men in the direction of France, always toward the ports of embarkation. Cantonments, except those on the Pacific coast, generally drew their men from westward. Draft trains usually moved toward the east. Thus were avoided millions of miles of duplicate travel.

During 1917 all selective service troops went primarily to the sixteen National Army cantonments. In 1918 this plan was changed, and selectives traveled directly to every camp, post, and station in the United States and in Alaska, Hawaii, and Porto Rico. This diversity of destination greatly complicated the problem of transporting the selectives.

The perfection attained by the troop-travel system is best exemplified by an occurrence of November 11, 1918, the day of the armistice; an achievement which, as the Provost Marshal General comments in his second annual report, "stands out as a marvel of efficiency." The Secretary of War had called for 250,000 new selectives to board the trains and travel to camp in the five-day period beginning November 11. The troop-movement office had received its two weeks' notice of this movement and had completed all arrangements. Schedules were printed and in the hands of the thousands who would need to consult them during the five days in question. As the hour of entrainment drew near, it became evident that Germany would accept the drastic armistice terms laid down by the Supreme Command. Still the War Department gave no sign that military preparation was to be stopped or retarded in anticipation of the event, lest this be interpreted in hostile quarters as a weakening of the national resolution. Word came that Germany had decided to sign the armistice, and still there came from the Department no response that could entail a slackening of activity. The Government remained unaffected by the spurious peace celebration which followed the groundless news dispatch that the war was over. It had been resolved that there should be no faltering of the morale by reason of any indications that victory was at hand. Early on the morning of November 11 the newspaper headlines proclaimed that the war was over. The four-hour difference in time made possible this announcement to breakfasting America. And not yet, with the unofficial news in everyone's possession, did the Government apply the brakes to the war machine. Washington was waiting for the official announcement to come from General Pershing. The draft boards and the railroad organizations had their orders to proceed with the entrainment of the 250,000, just as though victory were still a year away. On the morning of November 11 the draft trains started out and began picking up their little groups of selectives from station platforms crowded with people who were celebrating the dawn of peace.

In the troop-movement office the termination of the war was dramatic in the extreme. At 10.25 a.m. the order came over the private telephone wire directly from the Secretary of War to cancel the entrainment of selectives. Mr. Hodges himself, the chief of the office, took the message. Long distance telephone and telegraph circuits had been set up for hours. In a few minutes Hodges had either spoken personally over the telephone or had wired directly to the transportation agents at the six army department headquarters, ordering the movement of drafted men to cease. These officials relayed the orders to the state transportation officers and to the railroads. Inside of thirty minutes, draft contingents waiting on station platforms for trains that were never to arrive were notified that they could go home. Inside of the thirty minutes, every draft troop train on the rails was stopped, and many had turned backward and were distributing their passengers to the stations at which they had been picked up.

CHAPTER VI

INTERCAMP TRAVEL

SCARCELY had the first selectives been set down in the sixteen cantonments when the needs of the military service called for the transfer of large numbers of them from one camp to another. Thereafter for many weeks these transfers constituted a large part of the total military traffic. From the middle of October until the end of 1917, the so-called intercamp movement made up at least half the volume of troop travel within the United States. Intercamp travel continued, in fact, swelling or declining from time to time, until the armistice halted the march to France; but as the overseas movement expanded in 1918, it became proportionately a smaller and smaller part of the total travel.

Activating this heavy travel between camps were some of the most fundamental principles of military science. This travel was a bustling manifestation of the strategic architect's work of fabricating a modern army. To an outsider it might seem that the Government, if it were conscripting a field expedition, needed only to call men to a cantonment, organize a division then and there from the material at hand, train that division, ship it directly to the Port of Embarkation, and then repeat this procedure at every training center. But in practice the process was not so simple. In the first place a modern army, and particularly one sent to foreign soil thousands of miles from its sources of supply, must be a self-contained unit; it must be made up of specialist organizations of men specially trained. A properly proportioned army is by no means entirely a combat force. In fact, for every man in the fighting zone there must be, behind him and sustaining him, another man in the military service. The combatant troops are but the javelin head of the army; the haft is equally im-

Photo from General Motors Truck Company

MICHIGAN TOWN AND COUNTRY FOLK SAY FAREWELL AT STATION

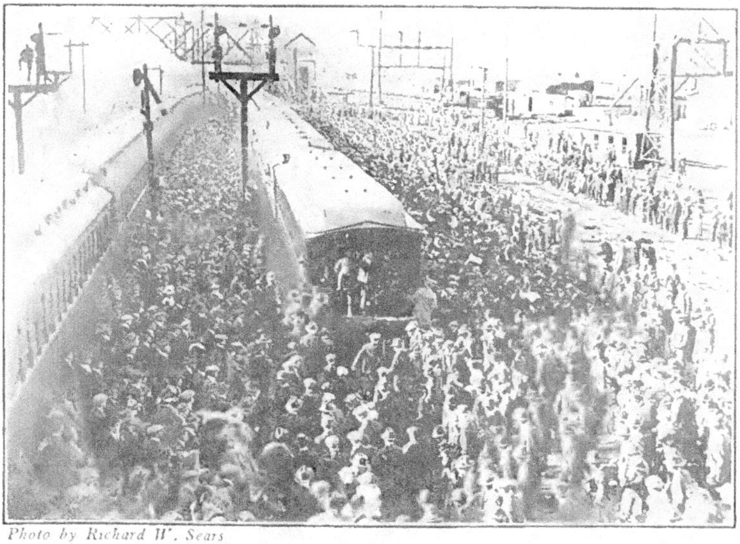

Photo by Richard W. Sears

A BOSTON CROWD AT DEPARTURE OF DRAFTED MEN

Photo by Winn & Traxham

THROUGH THE CANTONMENT *PERSONNEL* MILL

From The War College Collection

LINED UP FOR FIRST DRILL

portant, if less glorious, for along it move supplies, reinforcements, and human metal—the replacements—to repair the attrition where the thrust meets resistance. If we had placed 2,000,000 men on the front against Germany, it would have required another 2,000,000 men to maintain them there. Half a million of these last would have been combat troops organized in divisions and engaged either in training or in traveling forward as reinforcements. The million and a half would have been assisting corps troops of various sorts—service-of-supply troops, replacement troops, and the troops required in the operation of the great military establishment within the United States.

In a war so consuming as the recent one, the matter of providing replacement troops was not the least difficult of the problems with which the army authorities had to deal. When the shipments of troops to France had been properly regulated, a full quarter of those who embarked were replacement troops. One man in every four, then, sailed to take the place of some predecessor removed from active service by death, wounds, sickness, or other disability. In all, nearly a quarter of a million American soldiers crossed to France as replacements. And even the replacement troops were specialists. We did not herd men together indiscriminately, label them "replacements," and then ship them overseas, there to be assorted, filtered into units whose ranks had been depleted, and trained in those new surroundings. Our replacements first received their training here as infantry, artillery troops, machine gunners, and what not, and crossed the ocean in homogeneous units to go into the reservoir of men from which the combat divisions of the A. E. F. drew to fill up their files.

The building plan followed was to train at each of the sixteen cantonments one division at a time, made up of men from the district geographically tributary to that center. A division completely recruited, including its artillery, numbers only 27,000 men, whereas each of the cantonments had a housing capacity of 40,000 or more. The excess thousands were camp-maintenance troops, newly drafted men receiving first military training, and regiments of auxiliary corps or replacement troops undergoing special training in one branch or another.

Presently, in the evolution of the training scheme, each of the sixteen camps began to specialize in one or another of the principal army activities; for it was obviously inefficient for the curriculum of one camp to attempt to include all the specialties. For example, Camp Gordon in Georgia, Camp Lee in Virginia, Camp Pike in Arkansas, Camp MacArthur in Texas, and Camp Grant in Illinois specialized in the training of infantry troops, in addition to their primary occupation of whipping their resident divisions into shape. Most of the infantry replacement troops shipped during the latter months of the war came from these camps. Camp Hancock in Georgia, outside of its regular divisional instruction, specialized in the training of machine gunners. Camp Jackson in South Carolina and Camp Taylor in Kentucky trained field artillerymen. Camp Meade in Maryland specialized in the training of Signal Corps troops.

Then special camps sprang up exclusively for the specialized training of various corps troops. Camp Humphreys in Virginia became an enormous center for the training of engineer troops. Another engineer camp was Camp Forrest in Georgia. Near Jacksonville, Florida, the Quartermaster Corps established a great schooling center for its men: Camp Joseph E. Johnston. Camp Meigs in Washington, D. C., was another quartermaster training camp. The Motor Transport Corps, which during most of the war was part of the Quartermaster Corps, also provided special instruction for its men at Camps Johnston and Meigs. The Medical Corps set up troop-training schools at Fort Oglethorpe in Georgia and Fort Riley in Kansas. The Signal Corps maintained in New Jersey an exclusive training school called Camp Alfred J. Vail. Troops of the Coast Artillery received special training at Camp Eustis, Virginia. The new Tank Corps built Camp Polk in North Carolina and also operated a training camp at Gettysburg, Pennsylvania. The Chemical Warfare Service established a training camp at Lakehurst, New Jersey. And the Air Service maintained a large number of flying fields and training camps throughout the country.

In addition to these and other special camps, the Army met the exigencies of war by building up in this country an enormous operating establishment. There was a chain of general and special military hospitals, each one of which had to be manned with maintenance and operation troops. The great supply warehouses and bases in this country all required contingents of troops—some of them large contingents. It monopolized the energies of nearly 50,000 officers and men to conduct the manifold operations of the Port of Embarkation at New York, and another big force was required at Newport News. Troops were called out to operate some of the military manufacturing plants—especially those which involved danger, such as the toxic gas plants at Edgewood Arsenal, where military discipline alone could ensure the maximum of industrial safety, and where it was impossible to retain civilian workmen. The great military proving grounds, operated by enlisted men, each had to have a cantonment of considerable size. In addition, there were the coastal and interior forts of the United States to be maintained at strength; the military executive headquarters in Washington and elsewhere required soldiers as orderlies, chauffeurs, and the like; and there were hundreds of small military establishments and enterprises in the country, all needing the presence of troops in smaller or larger numbers.

From all this it is evident that for hundreds of thousands of draft selectives, the sixteen cantonments were only the first stop on a military journey which might or might not take them eventually to France, according to each man's special qualifications for service. As a fact, the cantonments became primarily great centers for the sorting of men into the supplies of human raw material which headquarters used in building a completely rounded army with a combat force in France and a sustaining force extending back and covering the United States.

When, after a final physical examination, prophylactic inoculations, and a brief quarantine, the new selectives reached their barracks, they were assigned to the camp depot brigade for the preliminary training in infantry drill and the manual of arms which each soldier of a well-disciplined army must undergo. It was during this interval that the men met the trade tests which constituted so remarkable a phase of the American military plan. These tests were devised by experts to include practically every known industrial and commercial activity. The soldier might claim too much or too little ability in any occupation, but an intelligent examiner could, with five or six of the test questions, rate him with extraordinary precision. Every soldier's special qualifications were thus catalogued promptly after his induction. In the card index in Washington reposed a detailed inventory of the total ability of the great American Army in every vocation in which even a single one of its nearly 4,000,000 members was qualified. Then entered the strategical architect, the Operations Branch of the General Staff, who took the data, sorted out the material, and assigned each man to an organization where his particular abilities—which might be, and often were, merely physical—best applied, and where he was most needed in order to maintain the steady flow toward France of the power that was to bring the foe to surrender.

Many of these special training camps were not in operation until 1918; the various cantonments had not as yet adopted specialties, and the system of training replacement troops in designated camps was not in operation until late in the spring of 1918; yet the intercamp movement began almost as soon as the first selectives reached the cantonments. This early translocation was due to another element, one which we have not yet considered. The divisional organization of the Army, as it stood in 1917 and 1918, was called into existence to meet the emergency of the war with Germany. The First Expeditionary Division was organized and its units were designated after April 6, 1917. The Second Division was organized entirely in France, its component units having crossed the ocean without divisional identity. In this country, meanwhile, divisions were being authorized, and divisional headquarters established at all of the thirty-two camps and cantonments. Several of these organizations were composed of Regulars. Seventeen were made up of National Guard troops. Regular divisions and National Guard divisions possessed, in the autumn of 1917, one characteristic in common: they were all skeletal in form, and waiting to be fleshed out with recruits. Practically every one of them had to be built up to strength with conscripted men.

A division with a trained nucleus could be pointed for foreign service much more quickly than one made up entirely of green troops. Thus it was obvious that the first divisions to be sent overseas must be the Regular Army and National Guard divisions. All of the National Guard divisions and some of the Regular Army divisions existed, at least structurally, and were in camp when the selectives were called from their homes. Consequently the immediate concern of the War Department was to take selectives from the cantonments and send them to fill the gaps in the organized and veteran ranks.

This necessity created a brisk intercamp travel as soon as the inducted men got fairly into their uniforms. In fact, the earliest intercamp trains began moving about October 10. At this time some of the National Guard trains were still entering the South. The troop-movement office found that it required delicate adjustment of schedules and dates to avoid overcrowding the rails and to prevent congestions at the gateways where north-bound intercamp troops and south-bound National Guard units might meet. The volume of intercamp travel multiplied forthwith. Between October 10, 1917, and the last day of December, more than 350 special trains carried more than 175,000 troops between camps. The total movement of troops between October 10 and the end of the year was slightly more than 325,000 passengers. Thus, in these ten or eleven weeks, camp-to-camp travel made considerably over half the total traffic. Some 500,000 passengers were carried on special troop trains from the outbreak of the war until

December 31, 1917. A clear third of this traffic consisted of intercamp travel confined to the last ten weeks of the year.

After the skeleton Regular and National Guard divisions had been filled up from the reservoirs of the National Army cantonments, and the National Army's own divisions were organized and in training, the special schools, now set up, began calling for soldier students by tens of thousands. Moreover, a new element had entered to complicate the matter. The need of replacement troops for the divisions in France had forced itself forward for official consideration. To find these replacements, the Army turned to the half-trained ranks of the National Army divisions, took men by thousands from those ranks, and dispatched them overseas as replacement troops, thus making it necessary to infuse fresh material into the plundered National Army divisions. The thinning of the National Army ranks continued until the late spring of 1918, when the plan of training special replacement troops at various designated camps went into effect.

After the earliest overseas divisions had been brought up to strength by the addition of men from the cantonments (and some of the National Guard divisions drew more than one-third of their strength from this source), the National Army had need to restore its own strength and balance; and this could be done only by the transfer of many troops. At the same time dozens of military stations and establishments, large and small, were springing into existence and demanding men. Such needs continued throughout the period of hostilities and made the transportation of troops between camps always a salient phase of inland traffic.

One more cause of extensive intercamp travel was a domestic condition peculiar to the United States. The negro problem created many a delicate situation in the construction of the Army. There were millions of negroes in the United States, a fact which forced the War Department into a policy of brigading them with whites. The general military policy was to compose each National Army division of troops drawn from the vicinity of its cantonment. Yet, when the selectives

Photo from American Red Cross

A TROOP TRAIN PASSES

From The War College Collection

A STOP IN A CALIFORNIA TOWN

Photo from American Red Cross

THE RED CROSS CANTEENS DREW NO COLOR LINES

Photo by Underwood & Underwood, N. Y.

AN ENTRAINMENT AT CAMP WADSWORTH

were called out, it was found that at certain southern cantonments this plan would create divisions in which there were more colored men than whites; whereas certain northern cantonments housed divisions in whose ranks were scarcely any negroes. Of this distribution was born the policy of giving to each National Army division a small percentage of negro troops, leaving each division preponderantly white. This policy implied the transportation of thousands of southern negroes to northern cantonments; and such withdrawals created vacancies in the southern divisions. Because it would have been unfair to the white population of the South to go deeper into the draft registration for selectives to fill such gaps, the War Department filled them by transferring white units from northern and western cantonments to those of the South.

The permitted ratio of colored men to whites in the divisions left unassigned a residue of thousands of negro troops for whom there was no room in existing organizations. To provide places for these men the Staff created two all-colored divisions, the Ninety-second and the Ninety-third. Neither was trained as a unit in a special camp. Their various components in training were scattered throughout the country. Each division assembled for the first time at the Port of Embarkation on its way overseas. During the training period the Ninety-second maintained headquarters at Camp Funston, Kansas. This division represented the colored people of the United States, for its troops had originated in practically every state of the Union. They trained in various cantonments, principally in the North. The Ninety-third was preponderantly southern in composition. Its headquarters were at Camp Stewart, near Newport News, Virginia, where the division assembled prior to embarkation.

Intercamp travel propounded an involved transportation problem in that its direction was not constant. All other principal military movements, except the travel of the National Guard to its training camps, was in tendency toward the ports. The intercamp travel, although the army organization attempted to give it a trend toward the ships and the ocean, remained crisscross. It was usually over long distances, and the intercamp trains were almost always made up entirely of sleeping cars.

With embarkation constantly on the increase, with the movement of selectives to training camps continuing steadily, and with intercamp travel at its maximum, military transportation reached, in the late autumn of 1917, an unprecedented volume. And then came the first of the holidays, Thanksgiving, to add to the burden the thousands of men who were traveling on furlough. It is estimated that considerably more than a quarter of a million soldiers received leave to go home at Thanksgiving and again at Christmas. It was at this time that the Government and the railroads asked the civilian population to stay off the passenger trains and to travel only on the most necessary business. Furloughed men at the holidays and on Sundays traveled on the regular trains, but their numbers drew heavily upon the available equipment of cars and locomotive power.

As army traffic grew, the skill of the transportation organization in charge expanded with experience. Special travel which, a few months earlier, would have demoralized the railroads and their war organization was handled with scientific and expert precision. By the time snow flew in the terrific winter of 1917-1918, the troop-movement system had attained a professional assurance which inspired in the Government the utmost confidence in its ability. The military authorities now knew that the system could meet any demand laid upon it.

We can get a picture of the teeming activity of those days by taking from the daily records of the troop-movement office a detailed account of military travel on December 25, 1917—the war Christmas. The Second, the Twenty-sixth, the Forty-second, and the Forty-first divisions were then moving through the Port of Embarkation at New York. Two trainloads of Second Division Infantry reached Camp Merritt, the port camp in New Jersey, early on Christmas morning. Two aëro squadrons, Nos. 167 and 168, which had traveled on a special train from San Antonio, Texas, reached Mineola, Long

Island, on the evening of Christmas Day, for early departure overseas. Another overseas unit, the 12th Field Artillery, reached Newport News, Virginia, on Christmas afternoon. The 303d Stevedore Regiment, with about 2,500 men, started from Newport News for Jersey City for embarkation, reaching Jersey City the morning after Christmas. Evacuation Hospital No. 3, which had organized and trained at Fort Oglethorpe, Georgia, started for Hoboken. The 30th Engineers left their camp in Washington, D. C., at 6.30 p.m. Christmas Day, on a special train for Hoboken. On that memorable Christmas sixteen special troop trains arrived at various destinations, fifteen other specials started out from various points, and thirteen trains spent the entire twenty-four hours moving along the rails. Approximately 20,000 military passengers were carried in special trains on that day—a fair average for that period of the war.

A considerable fraction of this traffic consisted in the movement of volunteer recruits, especially those who had enlisted for aviation. On Christmas Day two trainloads of aviation recruits which had traveled from San Francisco arrived at Waco, Texas. Two trainloads of aviation recruits from the barracks at Columbus, Ohio, arrived Christmas afternoon at the flying field at Rockford, Illinois. Another section of aëro recruits, made up of conscripted men from Camp Sherman, Ohio, reached Kelly Aviation Field, Texas, early Christmas morning. Another trainload of selectives from Camp Lewis at American Lake, Washington, arrived at the flying field at Rockford, Illinois, on that morning. A trainload of aviation recruits from Fort Logan discharged its passengers at Kelly Field, Texas. This train had followed to Kelly Field another trainload of recruits from Fort Slocum. A section bearing aviation recruits from Spokane, Washington, reached Waco, Texas, on Christmas morning. Another special train which reached its destination on that day bore 1,012 ordnance recruits from Fort Slocum to Camp Upton, Long Island. Three recruit trains left Fort Slocum on Christmas Day: one took artillery recruits to join the two divisions of Regulars then organizing at Camp Greene, North Carolina, another carried Engineer Corps recruits to Vancouver Barracks, and a third was loaded with recruits for Mercedes, Texas.

Nor is this all. On the rails that day rode a squadron of the 11th Cavalry in two trains *en route* from Fort Oglethorpe to San Diego, California; the 4th Engineers were on three trains bound from Vancouver Barracks to Camp Greene, North Carolina; the 1st Field Artillery rode in two trains *en route* to Fort Sill, Oklahoma, from San Francisco; recruits filled one train bound from Fort Logan to San Francisco, and another from Vancouver Barracks to Waco, Texas; the 8th Field Artillery traveled on three trains *en route* from Fort Sill to Camp Wheeler, at Macon, Georgia; engineer recruits were on a train bound for Camp Meade, Maryland, from San Francisco; and an intercamp train carried National Army troops from Camp Upton, Long Island, to Charlotte, North Carolina, there to join one of the regular divisions training at Camp Greene.

The military travel on our first war Christmas gives us a cross section of inland troop transportation at the end of 1917. The New Year of 1918 found the troop-movement office with a record of 2,850 special military trains operated for more than a million soldier passengers. These figures were exclusive of the transportation of nearly 400,000 selectives to their cantonments. By far the greater part of this travel had occurred in the last three months of 1917. The road to France was broadening. It pulsated to the tread of many thousands.

CHAPTER VII

THE SYSTEM AT WORK

FROM the 1st of January, 1918, until the armistice was declared there was never a day when fewer than three American Army divisions were traveling toward the ports of embarkation. Sometimes there were as many as five divisions simultaneously making their way toward the Atlantic seaboard. The columns were, to be sure, somewhat strung out. Certain of the divisions took several weeks, or even months, to move through the ports. But the average transit was shorter. Thirty days would be a generous measure of the average interval between the entraining of the first divisional units at a camp and the departure of the organization's last soldiers from American shores—a total which would require the major part of the activity to be compressed within a fortnight.

We may take as a typical instance of travel to the Port of Embarkation the journey of the Thirty-second Division, Michigan and Wisconsin National Guard troops for the most part, in January, 1918, from Camp MacArthur, near Waco, Texas, to Camp Merritt, New Jersey, whence it embarked on the transports. In making this journey, the entire division traveled over 1,900 miles. It moved on 61 special trains, on which 23,685 men were passengers. As was usual, the animals of the division moved considerably ahead of the troops. Two stock trains with the horses and mules left Waco on January 2 and 3 for Newport News, Virginia, at which port embarked most of the work animals of the A. E. F. Nearly 100 men rode on these freight trains to care for the stock. The troop movement itself began on January 10 and continued daily from the camp until January 23, when there was a break of a week.

On February 2 the activity began again; and on February 8 the last trainload of Thirty-second Division troops, except a company of infantry left behind to put the camp in shape and pick up any property overlooked in the departure, pulled away from Camp MacArthur.

Now let us analyze this entrainment in more detail. On January 10 a single train left the camp, bearing (1) the 107th Engineer Train, with its bulky impedimenta of camions and motorized shops, and (2) two sanitary squads which were to reach Camp Merritt in advance and put in shape the quarters assigned to the division's men. There were no departures on January 11; but on the 12th the 107th Field Signal Battalion entrained on one special train and the 107th Supply Train on another. January 13 witnessed the departure of two special freights, carrying the heavy equipment of the 107th Ammunition Train. Division headquarters, entraining on the 14th, led a close procession of troop trains across the country to Camp Merritt. The headquarters special was followed that same day by three trains carrying the troops of the 107th Regiment of Engineers. On January 15, four departures took from the camp such miscellaneous divisional units as the 120th Machine Gun Company, the headquarters of the 63d Brigade, the 107th Sanitary Train, and the headquarters of the 107th Regiment, together with the division's military police.

Then the four infantry regiments of the division began their entraining, the 125th Regiment on four specials January 16 and on three more the next day. The regimental Field Hospital was accommodated on two trains of January 17. On the 18th the 126th Regiment began its exodus from the camp, three specials starting on that day, two more on the 19th, and two on the 20th. Also on January 20, another section departed, bearing the headquarters of the 64th Brigade and the men of the 121st Machine Gun Company. January 21-23 encompassed the departure of the 127th Infantry on seven trains, three starting on the first day and two on each of the two succeeding days.

The 128th Infantry did not start until February 2, three

specials leaving the camp on that day and three on the next. On February 4, two companies of the 119th Machine Gun Regiment entrained on one special, and its third company, together with M Company of the 128th Infantry, on another. This departure left in the camp only the division's artillery and the single infantry unit, Company L of the 128th Regiment, which supervised the evacuation.

The Thirty-second Division's artillery, by rare exception, followed hard on the tail lights of the infantry trains ahead. (The artillery of our overseas divisions did not ordinarily embark with the other units: artillery and infantry usually traveled through different ports.) The freight of the 119th, 120th, and 121st F. A. left camp on a special train on February 5, preceded that day by five special troop trains carrying the headquarters of the 57th F. A. Brigade, the headquarters of the 120th F. A. Regiment, and the entire 121st F. A. On February 6, the headquarters of the 119th F. A. entrained on a special with the 107th Trench Mortar Battalion, the entire 120th F. A. Regiment entrained on three specials, and three companies of the 119th F. A. occupied another section—five specials on that day. On the 7th the rest of the 119th Regiment left camp on a special. On February 8, Camp MacArthur bade farewell to the Thirty-second Division when the last three trains departed, carrying the ambulance companies and field hospitals of the 127th and 128th Infantry Regiments and Company L of the 128th Infantry, the clean-up unit.

This was a typical divisional movement for embarkation, remarkable neither for its rapidity nor for its slowness. In a former chapter we have seen an example of great speed in troop transportation: the movement of the Seventy-ninth Division, less its artillery, from Camp Meade, Maryland, to shipside in Hoboken in two days. At one time during 1918 the executives in the troop-movement office compassed the unique achievement of an entire division, without artillery, moving along the rails at once. This was the Ninety-first, composed of selectives from the Northwest and Alaska, and trained at Camp Lewis, American Lake, Washington. Except for some miscellaneous advance units, the movement of the Ninety-first Division from Camp Lewis began on June 22. The troops left camp at the rate of seven trainloads a day, and the evening of June 28 found at the camp only the division's artillery. The last special had pulled away from American Lake before the first of the trains of June 22 had arrived at Camp Merritt.

The movement of a division at this rate and over such a distance appears even more extraordinary when one bears in mind the fact that a single division, properly accommodated aboard transcontinental trains, occupied half the total sleeping car equipment at the disposal of the military authorities. At the time when the Ninety-first Division was rolling eastward on half a hundred heavy trains, at least three other divisions were in process of moving to the ports of embarkation. Since sleeping cars and berths are inelastic, the simultaneous transportation of four or five army divisions in equipment with capacity for only two, verges on the impossible. Such feats were made practicable only by the most careful, systematic, and ingenious manipulation of the available rolling stock.

During the time of heaviest troop travel in 1918 there were approximately 6,000 Pullman sleepers in the United States. This equipment was none too large to accommodate the commercial demand; but the United States Railroad Administration, which in January, 1918, had taken over the operation of the American railways, managed, by cutting down the sleeping car accommodations on the regular trains and by discouraging all but the most urgent business travel on the part of civilians, to make available for troop movements about 1,500 Pullmans, leaving 4,500 sleeping cars for the use of the traveling public. Those 1,500 sleepers were the maximum wrung from the public during a time when every available transport was being loaded to capacity with American soldiers. Oftener, the sleeping cars at the disposal of the troop-movement office did not exceed 1,200. Since such a division as the Thirty-second, whose travel we have described, occupied 582 sleepers (and this division was by no means at full strength when it traveled to Camp Merritt), it is evident that the movement of four or five divisions at once taxed the resourcefulness of those at headquarters.

How was this car-equipment problem solved? By interlacing long-haul and short-haul movements. When four divisions were embarking at once, an observer at the port would have noted that two of them were coming in from far-away camps and two from camps near at hand. It is obvious that a division proceeding to New York from Camp Sherman in Ohio, an overnight ride, required a smaller amount of rolling stock than was necessary for the transportation of a division across the continent from Camp Kearney or Camp Lewis. A movement of the latter sort tied up equipment for at least two weeks—one week for the movement of the empty cars to the Far West and another for the loaded travel back again. The first trains of a divisional movement from Camp Sherman, Ohio, or Camp Lee, Virginia, could travel to the port, discharge their passengers and return their empty cars to camp before many additional trains had loaded.

The plan was, then, to sandwich short hauls between long hauls. Sleeping car equipment which had brought troops from the Pacific coast would be dispatched, immediately after unloading, to a camp in the East or Southeast. When it had returned to the port, it would be sent out to a greater distance. Overseas movements from the Pacific coast were few, compared with those from the interior sections of the country; hence the sleeping car shortage was never so serious actually as it was on paper. Up to the time the armistice was signed the railroads had hauled nearly 50,000 sleeping cars loaded with troops. It is evident that during the period of active hostilities every one of the 1,500 troop sleepers in the army equipment averaged between forty-five and fifty loaded trips. Since all the heavy travel came within the last twelve months of this period, each car averaged four loaded trips a month—a fact which testifies eloquently to the efficiency with which the equipment was handled.

The troops generally traveled in sleeping cars of the so-called tourist class. These cars materialized largely after war had come. In April, 1917, there were only some 400 tourist sleepers in the United States. They had been operated principally on the western trunk lines, for the convenience of landseekers going into the developing regions of the West. Travel in a tourist car, rating as second-class, was not so costly as in the standard Pullman, although the two cars are essentially alike except that the Pullman is somewhat the more luxurious in its appointments. To meet the war emergency, the Pullman Company converted 1,100 of its standard cars into tourist sleepers. The conversion amounted to little more than the removal of fine carpets and hangings; also, for the romantic appellations that once graced the sides of these sleepers, the painter's brush substituted prosaic serial numbers, thus answering the poet's rhetorical question to the effect that in the cost of railroad travel, at any rate, there may be something in a name. Each sleeping car furnished to troops carried a uniformed porter to minister to the wants of the passengers.

In the spring of 1917 the Pullman Company prepared for the coming military travel by equipping, as an experiment, four troop kitchen cars. The company did this by remodeling club smoking cars of the type used in the better trains. When ready for service, each kitchen car contained an equipment consisting of two large kitchen ranges, a warming oven, a long serving counter, an ice box to hold 2,200 pounds of meat, an extra large pantry, and additional water tanks with an air-compressor system for forcing water into them, so as to save the labor of watering the cars by hand. In each car there was also a commodious space for the dry storage of such food products as could be so kept. These cars proved that they could feed 800 men in an hour and a half, the food being taken through the train by squads of servers. Since the average troop train carried fewer than 450 men, the kitchen car was of ample capacity.

The experimental cars were built in the hope that the Government would try them out and order their general adoption. The Pullman Company offered a plan to equip fifty such cars, or enough to provide cooked food for the entire military transportation. This plan, however, was never adopted. The company certainly did not expect to operate these cars free of charge; and, if the war were to continue as long as the country expected it would, the use of a large equipment of them would run into a sizable bill. The railroads were required, on the other hand, to furnish baggage cars free for passenger trains. Consequently, the Army adopted the policy of equipping its free baggage cars as train kitchens by installing in them the ordinary army field ranges.

The Pullman Company did, to be sure, supply the troop movement with numerous cooking facilities. In addition to the four special cars, it put into service nine kitchen cars of a type which had already been in use. They contained fourteen sleeping sections, seven on each side. At the end was a large space outfitted as a kitchen, with a range and other facilities. These cars could feed from 200 to 250 men within a reasonable time. Also, the company went down into Mexico and took from the demoralized transportation system of that unhappy country three cars of a special type, known as hotel cars, which it had operated there. These contained only twelve sleeping sections, and considerable room was left for a well-equipped kitchen. The hotel car, also, could cook for 250 men. Thus the Pullman Company supplied, in all, sixteen special cooking cars for troops. This equipment was given constant use. The corporation supplied the cooks for these cars, but the troops themselves furnished the food supplies.

The commoner makeshift was the baggage-kitchen car. The Army devised the relatively safe method of installing field ranges in baggage cars by setting each range down on a thick layer of earth shoveled upon the floor of the car. Sometimes the company cooks were careless about following directions, and numerous fires in the baggage cars of troop trains were the result. The cooking in these cars was done by the cooks of the traveling units; so that the troops subsisted exactly as at camps and on long marches. One dusky magician of the stewpan, after being rescued from a charred and still burning baggage car, told his commanding officer that he had discovered in that fiery experience why his cabin in the South had twice burned down. Each field range was supposed to rest on a sheet of tin, which in turn was superimposed upon a layer of earth. This cook had reversed the procedure in the baggage car—no doubt following previous practice at home—by putting down the tin first, covering it with a layer of earth, and placing the bottomless stove on top of that. In a short time the car was on fire.

No refrigeration was provided on the baggage cars, and for fresh meats and vegetables the mess sergeants had to depend upon supplies purchased *en route*. Whenever an organization traveled without cooks, as casual companies, recruits, and intercamp troops often did, the soldiers carried rations with them, and the train commanders telegraphed ahead for hot coffee or other readily supplied foods. Such arrangements were particularly frequent during demobilization, when troops disembarking at New York traveled westward principally as casuals. The cookless trains fell under the special care of the Red Cross, whose terminal and junction refreshment stations rendered valuable assistance in keeping our traveling soldiers from going hungry.

The equipment division of the troop-movement office, originally called the Pullman division, manipulated all sleeping car and kitchen car equipment for troops and, after the early autumn of 1918, all coach equipment as well, the Railroad Administration having thrown all troop-train rolling stock into a single pool. The division, working in close coördination with the other branches of the office, was able to conduct distribution in time to meet any movement with the equipment which it required.

The troop-movement office was able to plan considerably ahead. The Operations Division of the General Staff aimed, in issuing its transportation orders, to keep a full month ahead of the actual movements. The principal officers in charge of transportation and its related activities in the Army met once a month and prepared a travel schedule for the following

Photo by Signal Corps

ARMY SLEEPING CAR. NOTE ABSENCE OF CURTAINS AND PARTITIONS

Photo by International Film Service

LOADING A BAGGAGE-KITCHEN CAR

Photo by Underwood & Underwood, N. Y.

A UNIT'S BAGGAGE INCLUDED ITS VEHICLES

Photo by Paul Thompson

LOADING A TROOP COACH

month. Having fairly definite knowledge of the tonnage that would be available at the ports of embarkation for the ensuing thirty days and of what the available tonnage would be on each day, the Operations Division prepared a table of transportation priorities, covering the order in which the units for overseas should come to the ports. The state of preparedness of various units for expeditionary service was known through reports which the Operations Division received from the Adjutant General. Consider, for a specific example, the port of New York. Just enough prepared troops were ordered to the port each month to fill the transports which would sail that month and leave in addition a reserve of 60,000 men at the two embarkation centers, Camp Mills and Camp Merritt. This reserve was always maintained, so that if there should be a transportation failure in the United States, due to storms, floods, earthquakes, or other unforeseen cataclysms, the transports coming in for troops would neither be delayed nor have to sail empty, but could continue to draw their human cargoes from the two embarkation camps. The Operations Division notified each prospective traveling unit of its place in the priorities, but ordered it to report its final readiness for entraining to the Transportation Service itself, which actually ordered the entrainment by date and hour.

The equipment division of the troop-movement section knew well in advance, then, to what regions it must divert its limited rolling stock. It accomplished its distributions by means of orders sent to the departmental traffic agents, who prepared the detailed orders for the railroad lines within their jurisdiction. The equipment division in Washington was the propelling force behind this distribution; the actual pull for the cars came from the general transportation agents at the camps where troops were to entrain. This double control resulted always in placing the equipment on time where it was needed. The efforts of the camp agent, it should be added, were sometimes quite as effective in securing cars as the original orders from Washington.

The troop-movement agent at each camp received periodical notification of what units would be expected to entrain at his camp. As the date of entraining drew near, he learned from the camp quartermaster at what hours the troops would be ready to go, how many would entrain, what personal or regimental baggage they would carry, what would be the number of their vehicles, cannon, and animals, and all other details which would enable him to estimate how much and what sorts of railway equipment would be required for each train. When he had these facts in hand, it was his duty to keep track of the railway equipment and see that it got to the camp on time. If he experienced difficulties, he could appeal to Washington, where resided the mandatory power.

When the equipment arrived, the camp agent inspected it. He could reject any that was unsatisfactory. Before any troops entrained, another inspection of equipment was conducted by the camp transportation agent jointly with the officer designated to be train commander. The commanders of traveling units had it within their province to say what units should travel on each train and in what order they should occupy the train from front to rear. As a rule, trains were loaded in accordance with that tenet of military doctrine which prescribes that tactical organizations shall not be broken on the march. When an organization moved on several trains, the policy was to keep these trains together, instead of splitting them up by inserting the trains of other organizations. Also, as a rule, a train was not divided between two units, but was occupied by one exclusively.

A troop train usually carried at least one freight car for baggage and equipment. In the late fall of 1917, Canada issued permission for the passage of American military trains across Dominion territory, and this favor gave us the advantage of using the three trunk lines that connect Chicago and Buffalo along the north shore of Lake Erie. On these and other northern routes, the winter weather, particularly that of 1917-1918, required the thorough heating of trains, which was impossible if freight cars were interposed between engine and passenger cars. The winter trains therefore usually carried their freight cars at the rear, except in states where that arrangement was forbidden by law. The presence of freight cars in troop trains necessarily slowed down the speed of travel, but this was not regarded as a disadvantage. In fact, troop trains were usually held down to an average running time of twenty miles an hour, partly for safety in operation and partly to make the transportation problem easier and prevent congestions which, if trains were moving into terminals at high speed, might in a few hours wax to unmanageable proportions.

The transportation agent at the camp, then, was in charge of the loading of all trains, jointly with the camp quartermaster and the train commanders. When a troop body of considerable size, such as a division, moved out of camp, the loading followed a systematic course. First, in box cars locked and sealed was loaded company property not needed in transit. Then came the loading of guns, artillery carriages, pontoons, wagons, ambulances, and other vehicles, in gondola cars or on flat cars. The forage for the animals was loaded in box cars. On the troop train itself was loaded all checkable baggage—arms, rations, and the like—for use *en route*. These properties were placed in the baggage and kitchen cars and kept under guard from beginning to end of the trip. The animals, last of the freight to be loaded, were put into stock cars, accompanied by soldiers to take care of them. Now all was ready for the troops themselves to entrain in coaches or sleepers.

Troop trains awaiting loads were backed upon convenient sidings a short time before the loading was to begin. The transportation agent, with a loading schedule in his hand, chalked on the side of each car, near the steps, at both ends, what organization should occupy that car and the number of men it would accommodate. On day coaches the troops were loaded three men to each double seat; on the sleepers they rode three to a section, two men occupying the lower berth at night and one man the upper. Troops marched to the railroad not more than fifteen minutes before their train's departure. Remarkable facility was attained in loading troop trains. At the embarkation camps, trains were frequently loaded within ten minutes; and the utmost allowance for loading, anywhere, was fifteen minutes.

The banner achievement in loading was that of the Eighth Division, Regulars, at Camp Fremont, California, in seven days, beginning October 18, 1918. According to the report by the entraining officer of this division, the first train left the camp at 9.00 a.m., October 18, followed by the others at regular one-and-one-half-hour intervals until 4.30 p.m. This procedure was repeated daily for seven days. The last train left promptly at 4.30 o'clock on October 24. The entraining officer reported: "The splendid coöperation of all units and train commanders and quartermasters made it possible to dispatch all of the forty-two trains on the minute specified, except two, one of which was held four minutes in order to remove baggage of sick officer taken off train just prior to departure, and the other of which was delayed five minutes to repair minor leaks in water connections. According to A. R. A. officials, the Eighth Division was the first division moved in the United States with 100-per-cent entrainment. Due perhaps partly to a competition inaugurated among train commanders, some very fast time was made in entraining; few trains after the second day required more than five minutes from the time of arrival of troops in area to the last man entrained." The officer cited specific instances of trains loaded in as brief a time as two minutes and fifteen seconds.

Soldiers are thirsty men. In addition to seeing to it that all drinking equipment was in good condition, cleaned, iced, and watered, the railroad agent at each camp had the duty of supplying additional drinking facilities. On the platform of every day coach carried on an American troop train, and on the platform of every second sleeping car, was a barrel filled with pure drinking water.

Each evening the transportation agent at every principal camp telegraphed in cipher to the central troop-movement office in Washington an account of railway departures from his station during the preceding twenty-four hours, the names of those organizations which were to move within the ensuing twenty-four hours, and the troop strength still left in camp. In September, 1917, orders went to the camp transportation agents to keep running diaries of the principal events occurring in camp, these to be forwarded to the troop-movement office in Washington at regular intervals. These narratives not only gave the central office a thorough knowledge of conditions at each camp, thus enabling them to act intelligently in matters of transportation affecting the camp, but it also put them in intimate touch with the spirit of the Army and built within the transportation organization an enthusiasm and morale that could scarcely have come from any other source. The executives of troop transportation, chained to their desks and telephones in Washington by a business which permitted not an instant's respite throughout the whole period of hostilities, became by this means almost as familiar with the military plant as if they had spent time traveling from camp to camp and inspecting conditions.

When a train started out from a camp loaded with troops, it followed a definite route to its destination. What road the train would traverse from a junction point on through the next stage of its journey was not left to chance or to the state of traffic. Every arrangement had been made in advance for the progress of that train from point of origin to terminal. This important department of operation was in the hands of another division of the troop-movement office, known as the routing section.

Proper routing was important, for more reasons than one, and it demanded in the routing officials a high degree of operating skill. The obvious benefit arising from good routing was the avoidance of congestion on the most heavily traveled roads. The chief concern of the routing officer was directed not so much to the more remote outlying railroads, which could handle their relatively infrequent troop trains without disturbance to regular traffic, as to the lines that converged upon the ports of embarkation, where, in early 1918, the principal concentrations of troops were beginning to occur. The routing official had to direct and distribute the flow of trains into these terminal channels so that the incoming movement would not choke any one of them.

In manipulating the troop traffic, the routing officer also had to keep a watchful eye upon the movement of freight toward the ocean terminals. Throughout the war, freight traffic was a source of much greater embarrassment to the railroads than troop traffic, although, as we shall see later on, the Government did eventually untangle the freight congestion.

Perhaps the most striking example of the coördinated handling of military freight and passengers occurred at the crossing of the Potomac River at Washington. Here, in the shadow of the Washington Monument, the whole north-and-south transportation system of the Middle and South Atlantic seaboards was pinched into two tracks. The system tapped territory as far west as New Orleans and as far south as Key West. This territory was the principal troop-training area in the United States. The many camps therein consumed an enormous volume of supplies, so that there was a heavy and continuous freight movement southward. The area was also a considerable munitions-producing section. Probably more troops traveled through Washington than through any other interior crossing or railroad center; and the transit of freight up and down the Atlantic seaboard section was always heavy. Into Washington, too, flowed coal from the West Virginia mining regions. Northward, between Washington and the port at New York, were located some of the principal munition and ship-building plants of the United States, as well as such huge purely military establishments as the Aberdeen Proving Grounds in Maryland and the Edgewood Chemical Warfare Arsenal near by. It is evident, therefore, that not only was the troop traffic heavy between Washington and New York, but that the freight traffic, too, was a thing of impressive proportions.

Now, between Washington and New York there are two railroad trunk lines, the Baltimore & Ohio and the Pennsylvania. The Pennsylvania is a four-track system over much of the route. The B. & O. is double-tracked over most of its right of way. Prior to 1917, the Pennsylvania, though it handled a great freight traffic, was predominantly a passenger railroad, its passenger trains considerably outnumbering those of the B. & O. between New York and Washington. The routing section of the troop-movement office simply interchanged these characteristics. It employed the B. & O. as its passenger road, thus freeing the four tracks of the Pennsylvania for the accommodation of the great freight traffic. When the embarkation movement was at its height, from twenty to twenty-four troop trains on the average were running daily from Washington to New York. Up to the number of eighteen, these were invariably routed over the B. & O., the Pennsylvania taking merely the excess. This example was typical of the war method of employing trackage for maximum service, regardless of its ownership and traditions.

The routing of military traffic in the West was affected by a historical consideration which touched the origin of certain of the railroads. When, on January 1, 1917, the transportation contract system gave way to a more scientific method of handling troops on the rails, the Government agreed (1) to use all railroads fairly and impartially in the event of war and (2) to pay for its passengers the lawful commercial fares with a five-per-cent deduction up to a specified gross maximum deduction. In the West there were certain railroads known as land-grant roads. To encourage the opening up of the West after the Civil War, the Government had offered the pioneer builders of railroads large tracts of land in the regions tapped by the lines. The land-grant arrangements had been sharply attacked in the decade prior to the World War, because the grants had become enormously valuable. Other pioneer rail construction had been financed by government bonds. There was, perhaps, a general feeling that the Government had been unduly generous in its encouragement of western railroads. However, in the charters of these land-grant and bond-aided roads were certain provisions ordinarily overlooked in times of peace—the Government's *quid pro quo.*

When the World War came upon us and we, as a nation, were faced with the expense of transporting troops by tens of millions, then these half-forgotten charter provisions became of the utmost practical importance, for they provided in perpetuity that the land-grant and bond-aided roads should haul the military passengers of the United States at greatly reduced rates, or even free of charge. The land-grant roads could charge, at most, only half fare for troops. Competing western lines, built later, had met these troop rates in their tariffs on file with the Interstate Commerce Commission, with only the exception that no competing line gave free transportation. In the agreement of January 1, 1917, all of these reduced rates were continued, and the further five-per-cent deduction was also allowed. The Government did not actually enforce the free-transportation obligations upon any of the land-grant roads, having mercy in view of the unprecedented military traffic in the recent war; but in some instances it did secure rates well under the half fares generally charged.

The land-grant-railroad situation inevitably had great significance for the routing office in Washington. Whenever it was possible, consistently with prompt dispatch, the office routed troop trains so as to take the greatest possible advantage of the land-grant fares. As a result, much of our troop transportation in the West in 1917-1919 was handled by the railroads at less than cost. By the time demobilization ended, the foresight of the empire builders of the mid-nineteenth century had been justified. Reckoning in the increment in the value of the Government's own lands in the West, we find the total cost-saving in troop transportation over the aided roads to have been a sum considerably greater than the value of the land grants at the time when the railroads were opening up the West for colonization. The war wiped the slate clean.

Often, too, money could be saved in the routing of troops by following the commercial ticketing arrangements of the railroads. The prudent traveler has sometimes noticed that

Photo by Paul Verkin

FOOD STORES FOR A JOURNEY

From The War College Collection

ARTILLERY TRAVELED APART FROM INFANTRY

Photo by Central News Photo Service, Inc.

DRAFT TROOPS IN COACH

Photo from American Red Cross

A WELCOME BREAK IN THE TEDIUM OF TRAVEL

he can ride more cheaply to a given destination upon one railroad and its connections than upon another. The routing section of the troop-movement office was aware of all these travel economies, and invariably it routed a train according to the lowest ticketing rate, except when it was necessary to sacrifice this policy to avoid railroad congestion.

The basis of fares during much of the earlier part of the war was approximately two cents a mile for parties of ten or more, less the five-per-cent and other deductions. In June, 1918, the Railroad Administration abolished party fares and carried troops on a basis of three cents a mile per capita, less the deductions. These deductions, combined with a thrifty policy in routing the trains, saved the country millions of dollars in its war transportation bill.

CHAPTER VIII

AT THE HEIGHT OF THE EFFORT

UP to the signing of the armistice the Army threw a veil of impenetrable secrecy about its travel. Even the railroad men secured only sufficient information to enable them to operate the trains intelligently. Originating orders for the travel of troops were usually oral rather than written. To prevent possible leaks in information through the commercial telephone exchanges, a direct and exclusive private line was installed between the troop-movement office and the War Department, and the orders which resulted in travel usually went across this line. The canteen service of the Red Cross received notice when a troop train was to stop where the men could be fed, but only briefly in advance of the actual arrival of the train; and when a military train was traveling to a port of embarkation, not even the Red Cross knew of its movements.

As an aid to this secrecy, the troop-movement office adopted an elaborate cipher code (the invention of Mr. J. Edwin Dempsey, a telegraphic cipher expert of Chicago), in which all of its confidential orders were sent. The code was published in a small leather-bound volume and placed in the hands of every man who held an executive post in the military transportation system.

Ciphers interest the inquisitive mind. The following jumble of words is a transcript of an actual Dempsey code message sent out from the troop-movement office during the early autumn of 1918:

"MAIN 89 WHITING EIGHTH REVERSED AMASSING FREMONT BARBETTA WHIST CLEAVING HEFTMAN RENNISH HEFTMAN RENOMED APATHY STRIVING

GARLAND SUMAGE GARRED SIMMERED FUMING SOBERING FULTUM SWARDED UTICA SPRIGGED GARISH SMILED SFIBRARONO."

A free translation of this message is as follows:

"Subject: Movement of the Eighth Division. Following is the routing for certain trains of the Eighth Division from Camp Fremont, California, to Camp Mills, Long Island. Movement to consist of three trains scheduled to leave October 24 and three trains October 25. Tourist sleeper equipment for all. Route of trains: Southern Pacific to Ogden, Union Pacific to Omaha, Illinois Central to Chicago, Michigan Central to Buffalo, West Shore to Utica, New York, Ontario & Western to New York City, Long Island Railroad to destination. Henceforth the code word for this movement will be Sfibrarono."

Now for the explanation, which will incidentally throw some light upon the system of handling troop traffic. The message begins with the number "Main 89," suggestive of the telephone directory. For the sake of both secrecy and brevity, every principal movement of troops received its serial number from the troop-movement office. "Main 89" was the serial number applied to the entire movement of the Eighth Division from its training camp in California, Camp Fremont. The movement in its entirety consisted of forty-two trains, but the whole was known to the train dispatcher simply as "Main 89." The sender of the message might have obtained even greater secrecy by calling the movement "Main Pardah," the word "Pardah" being the Dempsey code word for the numeral 89.

The third word of the message was one of the most important, for it indicated instantly to the recipient the character of the message. In the Dempsey code, "Whiting" meant that the following sentences referred to a route schedule. Other color words in the code meant messages of other purport. If the message were a report from a camp agent relating to the departure of troops from his camp, the serial number of the movement, which always headed the message, would be followed by the word "Black." Every person concerned who handled this message would call it a "Black report," and would know without deciphering it that it related to trains starting out. A "Gold report" related to the delivery of a troop train from one railroad to another at a junction. A "Yellow report" was the report of an arrival at destination. "Whiting" meant the routing of a movement not yet begun.

Here are the six code words which follow "Whiting," with their translations:

Eighth	Eighth
Reversed	Division
Amassing	Camp
Fremont	Fremont
Barbetta	California
Whist	Camp Mills, Long Island

Then came "Cleaving," important because it was a key word. An expert in the code knew at once that "Cleaving" embodied the idea of the make-up and the departure of the movement. It may be roughly translated, "Movement scheduled to leave as follows."

Heftman	Three passenger trains
Rennish	October 24
Heftman	Three passenger trains
Renomed	October 25
Apathy	Tourist sleepers will be provided
Striving	Southern Pacific Railroad
Garland	Ogden, Utah
Sumage	Union Pacific Railroad
Garred	Omaha, Nebraska
Simmered	Illinois Central Railroad
Fuming	Chicago
Sobering	Michigan Central
Fultum	Buffalo, New York
Swarded	West Shore Railroad
Utica	Utica, New York
Sprigged	New York, Ontario & Western Railroad
Garish	New York City
Smiled	Long Island Railroad

Finally came the word "Sfibrarono." This cryptogram was a key word invented by the sender especially for that message. It was what was known in the transportation system as the route word, and it meant that henceforth all telegraphed messages relating to the movement of these six trains of the Eighth Division over the route prescribed in their "Whiting" message must be identified at the beginning of each telegram by the word "Sfibrarono." A cipher report of the delivery of one of these trains from, say, the Union Pacific to the Illinois Central would read, "Sfibrarono Gold," etc. When the six trains had reached their destination, the word "Sfibrarono" would drop out of existence.

The transportation of the Army was an almost sheer addition to the burden of normal railroad traffic. Upon the railroad telegraph system it placed an excess load hardly less staggering. The ingenious Dempsey code was a great aid in keeping the military-traffic-wire business to a minimum. The message quoted above contains thirty words. To have telegraphed the same information in plain English would have required nearly one hundred. When the armistice was signed and there was no longer the need for secrecy in military movements, but when there was still a great army to be demobilized and taken back to its homes, the code was continued in use for the sake of its succinctness; and it remained as an efficient transportation tool until, with the demobilization of the First Division (the last to return from France), the great episode in military travel came to an end.

It was the Eighth Division whose rapid entrainment at Camp Fremont was commended by the authorities. The entire division moved out in seven days, beginning October 18. Each day six trains departed punctually on schedule. All available routes were so utilized in this movement as to prevent congestion on any one of them. The average run from Camp Fremont to Camp Mills was 3,444 miles long. The journey of the Eighth was still more memorable because it occurred in the height of the influenza epidemic. Many a man was taken from the trains of the Eighth Division to hospitals along the route, and many a one who entrained at Camp Fremont was dead before his comrades saw the Manhattan skyline.

The avalanche of American troops upon the ports began suddenly. To hold the German drive in the spring of 1918, the British and French armies exhausted their troops at a ruinous rate. The American transport fleet had already grown to considerable proportions when the arrangement was struck whereby the British offered to turn over a large portion of their passenger-carrying marine if we would supply the men to fill the ships. The British agreed to transport overseas six American divisions. The question became merely whether the inland transportation system of the United States could move the troops to the seaboard in such numbers as to load the ships. It could. Not a transport, British or American, was delayed for want of passengers; not one left our shores unfilled. Up to the 1st of May, eight American divisions had been transported across the ocean, in part or as wholes. The month of May was to witness the ferrying of nine others—more than we had transported to France during all the previous months of the war. June saw seven more divisions embark; July six others.

All this meant tremendous travel on the railroads. In January, before the great embarkation was under way, 398 special trains had been operated for 137,093 passengers. The heaviest day was January 22, when 26 specials moved with 9,459 men. The intercamp movements had greatly fallen off by this time: during January they required only 13 special trains. In February embarkations were not heavy; but in March the Third Division, from Camp Greene, North Carolina, and the Seventy-seventh, New York City's own National Army division, which trained at Camp Upton, were boarding the transports. Train movements in February entailed the operation of 499 train sections for 151,987 passengers; in March 548 sections took 224,246 passengers. April saw a slight increase in travel—696 sections with 297,713 passengers. The great jump came in May. The British tonnage was now at the port. To supply the new demands the transportation system operated 1,063 special trains and carried considerably more than half a million soldiers. This was handling in a single month a volume of traffic almost equal to half the total military travel during the entire calendar year 1917.

To follow this travel in any detail would be monotonous. All we can undertake is to view the totals and in imagination translate them into terms of the physical activity on our rails. In June, over 1,200 sections moved, carrying more than 500,000 troops. On the 11th of June 43 special troop trains started out from various points with more than 30,000 passengers. Each one of these trains—and this was but a typical day of that period—required of the organization in charge that minute attention to detail which has been described. The crest of the load came in July, when 1,277 special trains, carrying uniformed and organized troops, traveled in the military movement. The July passengers numbered 672,266—a figure which excludes soldiers who traveled on regular trains and also the 401,000 drafted men inducted into the Army during that month. In all, the total military passenger traffic of July, 1918, was 1,147,013 men, a greater traffic than was carried during the whole of 1917.

The day of heaviest travel in July was the 13th, when 77 special trains departed, carrying over 41,000 troops. These figures indicate departures only: they take no account of trains which had left their points of origin before the 13th, but had not yet reached their destinations. It is conservative to estimate that at least 200 special troop trains were running on July 13, 1918. The train departures of that day, shown in tabular detail, were as follows:

No. of Trains	*Identity of Troops*	*Point of Departure*	*Destination*
8	Eighty-fifth Division Infantry	Camp Custer, Mich.	Camp Mills, L. I. (Embarkation)
3	301st Engineers	Camp Devens, Mass.	Bush Terminal, Brooklyn (Embarkation)
2	Marines	Miami, Fla.	Philadelphia (Embarkation)
1	Coast Artillery recruits	Jackson Barracks, La.	Fort Williams, Me.
1	Medical Corps recruits	Jackson Barracks, La.	Allentown, Pa.
1	Recruits	Columbus Barracks, Ohio	Camp Hancock, Ga.
1	Casuals	Camp Funston, Kan.	Fort Crook, Neb.
1	Unassigned	Minneapolis, Minn.	Fort Benj. Harrison, Ind.
5	Eighty-first Division Infantry	Camp Sevier, S. C.	Camp Upton, L. I. (Embarkation)
2	Replacements	Camp Gordon, Ga.	Camp Merritt, N. J. (Embarkation)
1	Unassigned	Fort Totten, N. Y.	Camp Eustis, Va. (Embarkation)
2	519th Service Battalion	Camp Devens, Mass.	Hoboken, N. J. (Embarkation)
2	Intercamp troops	Camp Lewis, Wash.	Camp Kearney, Cal.
3	Field Artillery of Seventy-ninth Division	Camp Meade, Md.	Philadelphia, Pa. (Embarkation)
1	Seventy-ninth Division (freight)	Camp Meade, Md.	Newport News, Va. (Embarkation)
26	15,926 overseas troops	Camp Mills, L. I.	Long Island City, L. I. (Embarkation)
1	Unassigned troops	Minneapolis, Minn.	Camp Dodge, Ia.
1	57th Engineers	Camp Meade, Md.	Philadelphia, Pa. (Embarkation)
2	Stevedore troops	Camp Lee, Va.	Newport News, Va. (Embarkation)
1	Engineer replacements	Camp Humphreys, Va.	Philadelphia, Pa. (Embarkation)
2	Engineer replacements	Washington, D. C.	Philadelphia, Pa. (Embarkation)
1	Student officers	Camp Sherman, O.	Camp Hancock, Ga.
1	Veterinary troops	Camp Travis, Tex.	Camp Lee, Va.
1	Unassigned	Fort Sam Houston, Tex.	Camp McArthur, Tex.
7	Overseas Engineers and Artillery	Camp Upton, L. I.	Long Island City, L. I. (Embarkation)

The procession of trains from Camp Mills to Long Island City on July 13 was a typical movement, duplicated time and time again during the spring and summer of 1918. Troops from Camp Merritt moved to the piers on ferry-boats down the Hudson River; but Camp Mills, having no water connection, sent her embarkation troops by the Long Island Railroad to the piers in the East River, where they marched upon ferry-boats. These train movements from Camp Mills usually began at the early summer dawn, the trains leaving the camp at twenty-minute intervals until about noon. Some of the trains, with as many as 1,000 passengers aboard, were unusually heavy. The railway journey to the ferry lasted about forty minutes. The ferry-boats were in the slips to take up the men as rapidly as the trains discharged them.

August was another peak month. The special movement involved over 1,300 trains, and the total of military passengers, including nearly 300,000 selectives, again went above the million mark. In September the movement declined somewhat, but, even so, more than 1,000 special trains were operated, and nearly 500,000 men were carried, not counting the inductives transported that month.

In September another burden was placed on the capable back of the troop-movement office: it was charged with the duty of transporting civilian labor supplied by the United States Employment Service. In this month the organization operated ten special trains for the movement of labor and, in addition, transported over 75,000 laborers in special cars attached to regular trains. The movement was principally from the Far West, where there was an excess of labor, to such great government institutions as the powder plants at Nitro, West Virginia, and Nashville, Tennessee, where there were labor shortages. In October the movement of laborers required the operation of twenty-three special trains; in November, of fourteen. The total of civilian laborers transported on special trains up to the date of the armistice was approximately 50,000.

As if to compensate for this additional traffic, the movement of selectives from their homes to camp somewhat subsided in volume after August 1, 1918. By that date the calls for troops had virtually exhausted the physically perfect and unencumbered men between the ages of twenty-one and thirty-one, and the Government was then preparing for the registration of all men between eighteen and forty-five. Meanwhile, the authorities did not wish to disturb industry by dipping down into the deferred classes of the first registration; and the Provost Marshal General accordingly slackened the induction of selectives until the new registration could be held in September. A more important consideration was that at this time the crops were being harvested. The War Department was unwilling to withdraw laborers from such indispensable work. In fact, draft calls would have been suspended altogether during this period except for the extreme necessity of keeping up the flow of men to France. The greatest of all interferences with the transportation of selectives was the influenza epidemic which set in over the country about October 1, 1918. Because of epidemic conditions, draft calls for entire states were frequently suspended, and as a result the transportation diminished in volume.

In October the regular movements of troops entailed the operation of 869 special trains for approximately 320,000 passengers. During the first eleven days of November, 241 specials carried about 100,000 troops. Nor did military movements then immediately stop: from Armistice Day to the end of November the Government operated 217 special troop trains for approximately 95,000 soldiers.

The great offensive undertaking which had begun in earnest May 1, 1917, and come to a close on the 11th of November, 1918, had required the operation of 11,972 trains over 10,000,000 train-miles, and had carried 5,051,608 passengers. In addition to these, nearly 1,500,000 troops had ridden on regular trains. Nor do these figures include the transportation of selectives, who numbered about 2,300,000. These round out a total traffic figure of over 8,500,000 military passengers transported during the forward movement. These men were

Photo by Bate Studio

VOLUNTEER LABORERS LEAVING PRESCOTT, ARIZ., FOR NITRO POWDER PLANT, W. VA.

Photo by Norman A. Burke

A FAMILIAR STATION SCENE IN 1918

Photo from American Red Cross

A RED CROSS RAILWAY CANTEEN

Photo from American Red Cross

MAIL FACILITIES *EN ROUTE*

carried with an equipment which, at its maximum, did not exceed 1,500 sleeping cars, 2,500 coaches, and 500 baggage and express cars. The average military train had 12 cars, carried 421 men, and traveled 785 miles in 38 hours at the rate of 21 miles an hour.

The men who rode on these trains rode with a degree of safety unknown to the commercial passenger. The wrecks and accidents were blessedly few. Whether or no because of the secrecy with which the travel was invested, the passengers were safe from the machinations of enemy agents. The country sometimes heard rumors of troop trains wrecked and of plotters and spies taken from trains and summarily executed beside the track, but these things never occurred. In all there were 17 train accidents, in which 41 men were killed and 338 injured. The worst wreck occurred on September 17, 1918, near Mansfield, Missouri, when a special train carrying selectives of the September automatic replacement draft went head-on into a freight train on the St. Louis & San Francisco Railroad. The engineer of the troop train failed to observe a block signal set against him. There were 507 selectives on the train, of whom 12 were killed and 37 injured. Three trainmen also lost their lives.

On the morning of November 11, the handful of executives in the troop-movement office in Washington, who had managed successfully the hugest railroad transportation of soldiers over the longest distance ever attempted, looked up from their work, which now for the first time did not have to be incessantly fought down lest it overwhelm them, and began to take stock of what had been done. The thing had been accomplished; the war was won; and their part in the victory had been a vital one.

That of which this little band of practical railroad men were perhaps proudest was an incident that occurred in the latter part of August, 1918, when the military travel was heaviest. They had made all arrangements, dispatched the equipment from a dozen points, and completed the routes and schedules for the transfer of 15,000 artillerymen from Camps

Taylor (Kentucky), Robinson (Wisconsin), and Doniphan (Oklahoma), to the seaboard for embarkation. The trains of empty cars were already converging at these points, when the War Department called the liaison officer into consultation and advised him that the urgent overseas need was for infantry. Consequently, the Department had decided to cancel the sailing of the artillery and to substitute on the same ships 17,000 infantry troops from Camp Sherman (Ohio), Camp Dodge (Iowa), and Camp Funston (Kansas). The ships were even then approaching port; some of them, in fact, had arrived and were unloading. Could the troop-movement men accomplish this substitution without a break in the flow of troops to the ports?

It was a question soon answered in results. Within twenty-four hours, messages had caught every item of the equipment *en route* to Camps Taylor, Robinson, and Doniphan, and had diverted it to Camps Sherman, Dodge, and Funston. Some units of the designated infantry were on trains in this same equipment two days after the War Department had ordered the change. Only an organization proficient and flexible in the extreme, as well as in instant control of all the transportation facilities, could have handled so quickly this difficult railroading problem. By such examples of extraordinary operation, the troop-movement office, an organization of civilian volunteers, continually justified its existence as an integral part of the military traffic machine.

CHAPTER IX

THE WAR FREIGHT PROBLEM

THE railroad movement of troops during the war was great in volume and spectacular in accomplishment; but it was, after all, the less difficult branch of military transportation. Wherever a soldier went, over land or sea, several tons of supplies each year—his food, his clothing, his weapons and ammunition, materials for his shelter, and thousands of other things, a range of articles which embraced virtually all the products of industry, all the things which kept him fit and efficient as a fighter—must follow him. The railroad transportation of these material supplies was a much harder problem than moving the troops. The Army quickly found that it could rely upon the railroads themselves—that is, upon the troop-movement office of the American Railway Association, which later became the troop-movement section of the United States Railroad Administration—to manage all the troublesome details of the travel of army personnel; but when it came to the movement of the supplies, nothing short of a stern, ungloved military dictatorship of freight transportation was sufficient to bring order out of a perilous confusion and save the nation from a disaster.

The Army was compelled, then, to take into its own hands the transportation of its supplies. It adopted this drastic measure only after the failure of the railroads had brought the military supply system to the very brink of demoralization. The rescue of American freight traffic from the chaos which almost engulfed it late in 1917 was one of the stirring episodes of our participation in the World War. The transportation of troops has, of course, the more direct human appeal; yet there is, in the prosaic progress of the military freight train, rattling and clanking along toward its destination, another and not less intense drama of human struggle and final triumph over difficulties for the reader's imagination to compass. To comprehend this drama, one must first know something of the traffic problem that confronted the War Department at the beginning of 1918.

Return in fancy to the days of crinoline and candles, post-chaises and clipper ships; the days when beaver-hatted dignitaries gathered ceremonially to drive the last silver spike which celebrated the completion of that astounding novelty, a new steam railroad. The locomotive wore a decorative grille to top off its smokestack; the fireman balanced himself on a precarious platform; and (if we are to believe the ancient woodcuts) a train, whenever it passed through a town, was accompanied by spidery little boys running beside the track and waving velvet caps with ribbons hanging down behind. In that transitional time the railroad builder put freight cars on his track for use there, and there only.

As long as the road remained isolated, unconnected with others, well and good; but when system after system sprang up and touched each other, until eventually a joined network of lines covered the entire land, then freight cars began obeying, not company regulations, but economic laws. A car was loaded with freight for a consignee beyond the limit of the railroad's trackage. At that limit, it was obviously laborious and costly to unload the car and transfer its freight to a car belonging to the connecting carrier. Instead of that, the inevitable recourse was to switch the car to the connecting line and haul it along toward its destination, transferring it to as many railroads as were necessary to bring it to the consignee. There, when it was unloaded, it was not to be sent back empty to the owning line; for that, too, would have entailed useless effort. Instead, the last railroad to receive the car used it as if it were its own, delivering it to a shipper along its line for a load. The car's destination this time might be even farther away from its ownership. Although, of late years, roads have made strenuous efforts to corral their absent equipment, freight rolling stock has tended to remain the bondservant of business, knowing no owner. Sometimes it is years before a wandering car chances to find itself upon home rails.

Thus the original supplies of cars became as thoroughly mixed as lottery numbers in a jar. The circumstances of inland commerce threw the entire freight rolling stock of the country into a pool from which any road or shipper could draw. Freight cars became a common commodity, like money, circulating from hand to hand. Under such a system, abuses sprang up. Certain entire railroad systems built no rolling stock at all (except locomotives), and relied entirely upon the cars of other lines. The result was that roads with heavy investments in freight cars frequently found themselves with insufficient rolling stock to handle their traffic, whereas other lines with a scant equipment of freight cars or, actually, no cars at all, were holding excess numbers of cars for the benefit of customers along their rails. Shippers had come to use railroad equipment as a universal commodity, held in only the vaguest bonds of private ownership. A broker who had a consignment of goods approaching a market might hold his stock aboard its cars while he waited for better prices. In times of car stringency, shippers who could secure rolling stock retained it on their sidings as insurance against future requirements.

The railroads presently took steps to curtail such abuses. The larger owners of rolling stock attempted to keep track of their equipment by elaborate systems of index cards and wall charts. But these owners sometimes found themselves powerless to draw in their cars, even after they had located them. There arose joint railroad committees for relocation of cars; there arose, also, the current system of interline car rules and regulations. Further to prohibit such unwarranted uses of equipment, the railroads prescribed penalties, demurrage charges. Such measures, however, could not correct all the evils; and it became a function of government regulation to guarantee the equitable distribution and use of freight rolling stock. The Interstate Commerce Commission is chiefly known to the general public for its activities in regulating freight rates and passenger fares; nevertheless, a great part of its work is concerned with the operation of the car equipment of the country.

Modern railroading possesses, besides its pool of cars, certain terminals where freight converges, and certain lines operating out of these terminals. Some of the lines are strong and active; others are smaller and not so aggressive. The great lines, waxing in strength, established business connections, with mutual understandings and agreements, in such wise that an outlying connecting road, lacking specific instructions from a shipper to the contrary, would route its cars over the strong and aggressive allied road, which would benefit financially by sharing the freight charges collected. The strong roads of the land have ramified their traffic-soliciting departments until the agents of those departments are in every principal town and city in the United States, and even in foreign countries. The New York Central Railroad proper connects New York and Chicago along the route of the Great Lakes: yet the New York Central maintains, for example, in Louisiana, a thousand miles from its tracks, freight agents who go to the Louisiana lumber shippers to persuade them to route their eastern business over the New York Central lines.

The solicitation of business by terminal lines has had a strong bearing on the development of cities in the United States. Such development has been striking on our Atlantic seaboard. Norfolk, Virginia, for instance, has as convenient, as well sheltered, and as commodious a harbor as New York; but the strong, aggressive lines have been those terminating in New York. The result is that New York handles the great bulk of Atlantic export traffic from the United States, whereas the excellent facilities of such harbors as Norfolk, Boston, Philadelphia, Baltimore, and Charleston have experienced no such development. This evolution is perhaps not disadvantageous during normal periods, but in times of transportation stringency it has resulted in a shockingly inefficient use of our traffic facilities. At times when the ocean terminals of certain lines have been literally glutted with traffic and the resultant delay in transportation has caused nation-wide suffering, other terminals—sometimes those of roads terminating at New York itself—have been working at less than their capacity.

Another inefficiency brought about by railroad competition for business was the operation of cars over much longer routes than were necessary. As an illustration, the Chicago, Rock Island & Pacific Railroad handles freight between Chicago and Little Rock, Arkansas. The distance over this route between these two points is 1,277 miles. The distance between Chicago and Little Rock *via* the Chicago & Alton Railroad and the Missouri Pacific is 633 miles. In other words, freight traveling on the Rock Island from Little Rock to Chicago actually goes more than twice the distance of the short route between these points. This is only one of dozens of instances in which competition has resulted in the inefficient employment of power and rolling stock. These inefficiencies apparently had little effect on the political economy of the country in normal times—although, of course, in the last analysis the general public pays for every pound of coal expended in waste motion and for every day's wear and tear upon a car traveling more miles than it should to reach a given destination. But when we went into a war that demanded the concentration of every resource of the United States to the single end of victory, then these inefficiencies became a matter of paramount national concern.

The fact was that American rail transportation was an industrial tool too unwieldy for us to handle as we were attempting to handle it. It was like an automobile with four or five operators—one man at the steering wheel, another at the gas, a third at the gears, and so on—safe enough so long as the machine could run along a broad, level beach with no external interference to the concerted attention, but tolerably certain to come to grief in heavy traffic. The railroad system was operated by its several hundred independent units, each a competitor of the others, and each actuated by a necessary self-interest to take what it could, especially in times of stress. All worked well normally; but in an emergency America presented the anomalous picture of a nation equipped with the finest and most complete transportation system in the world, actually suffering because of the failure of that system to function efficiently.

For years and decades this has been the state of affairs. A man has a long memory if he can recollect a time when the movement of the grain crop has not caused a car shortage embarrassing to other great national activities. Since America became industrially supreme there has been scarcely a winter in which it has not been difficult to secure an adequate supply of coal, thanks again to the inefficient administration of traffic. Paper makers have grown used to the regular annual shortages of pulp wood, due to the failure of transportation to send a sufficient number of cars into the logging areas. The chemical industry has suffered for phosphate rock and for sulphur, usually at regular annual periods. Great crops, such as the Pacific coast's perishable fruit crop, have seldom moved out as rapidly as the fields and orchards could supply the lading. Every person, whether he knew it or not, has experienced some personal inconvenience as the result of these periodical failures of railroad transportation. The phrase "car shortage" is familiar to all. Yet there have always been enough cars for our transportation needs. The actual shortage was a shortage not of cars, but of efficiency in the control of them. If American transportation forged ahead and kept pace with industry, it was only by main strength and awkwardness.

Then, in 1914, when the Great War broke out in Europe, America, always theretofore essentially an importing country, suddenly became the world's exporter preëminent. We became all at once the munitioning country for the European Allies. The manufacture of munitions bounded into prominence until it had become a dominant part of American industry. The products of this manufacture had to cross the ocean, and that meant that they had to concentrate at the seaboard over American rails. When this concentration was attempted on the system of operating American railroads through their independent units, the result was trouble.

By early December, 1916, the worst freight car congestion in the history of American railroads had clogged and choked the port of New York. For miles upon miles out of New York in every inland direction, cars were stacked up along the sidings and switches. Rail traffic suffered a partial paralysis. Congestions similar, though smaller in degree, were tying up traffic in all of the principal producing areas of the country. Railroads which were heavy originators of freight traffic seized and held every bit of equipment that came upon their lines, and the cars available for the occasional shipper virtually disappeared from the rails. America's magnificent equipment of nearly 2,500,000 freight cars, enough, if economically employed, to move a great part of the traffic of the whole world, was tangled in an almost inextricable knot.

The situation was described by the Hon. C. C. McChord, a member of the Interstate Commerce Commission, who conducted the investigation of the car supply at that time, in these words: "In some territories the railroads have furnished but a small part of the cars necessary for the transportation of staple articles of commerce, such as coal, grain, lumber, fruits and vegetables. In consequence mills have shut down, prices have advanced, perishable articles of great value have been destroyed, and hundreds of carloads of food products have been delayed in reaching their natural markets. In other territories there have been so many cars on the lines of the carriers and in their terminals that transportation service has been thrown into unprecedented confusion, long delays in transit have been the rule rather than the exception, and the operation of established industrial activities has been made uncertain and difficult."*

All this meant suffering, men out of work, people doing without the things they should have had; and it meant waste, loss of perishable foods, damage to goods exposed overlong to frost, rain, and sun. More than that, it meant the humiliation of a nation which prided itself upon its industrial genius, yet was unable to handle the most indispensable tool of its indus-

* I. C. C. No. 9284. Car Supply Investigation. January 18, 1917.

try—its railroad transportation. The fault lay, not in the men in charge of transportation—for, as they were to show later, their ability fully measured up to the requirements—but at the door of the system of control. The system itself was basically inefficient.

The railroads tried valiantly to untangle the jam. Conditions were still bad when, in April, war was declared—a crisis which made it of paramount importance to the very safety of the nation to bring about an improvement. The railroads appointed a Committee of Five to act as the central administrative agency of an attempt to operate the railway system of the United States as a unit. This committee consisted of Messrs. Samuel Rea, president of the Pennsylvania Railroad, chairman; Julius Kruttschnitt, president of the Southern Pacific; Hale Holden, president of the Burlington; Howard Elliott, president of the New Haven; and Fairfax Harrison, president of the Southern Railroad. The committee created a special car-service commission, all of its members traffic experts, to supervise the adequate distribution and movement of rolling stock. The railroads of various districts organized additional committees to supervise local compliance with the orders of the Committee of Five. The most important of these district operating committees was that at Pittsburg. This body was organized by the eastern railroads—those which fed into the principal Atlantic ports. The Pittsburg Committee attempted, by fixing preferences and priorities, to give our export transportation that unified order without which, under the load which American belligerency was about to place upon it, it must certainly fall.

All these expedients, remedial to a certain degree, were not sufficient to cope with the situation. The railroads found their hands tied by federal laws which prevented combinations and the pooling of their facilities. The orders of the Committee of Five were, at best, but advisory; they lacked the mandatory power which could have made them, in the true sense, executive. These reforms and the advent of mild weather in the spring of 1917 brought some alleviation. But the railroads were still staggering under their load of freight when there descended upon the tracks a crushing weight of building materials to be hauled to the sites of the new military training camps. At the same time, new munition plants to meet the Government's needs were springing up in all sections of the country, and the first traffic in America's own munitions was beginning to flow. The War Industries Board in Washington and the various producing bureaus attempted to put in force some sort of priority in shipments. In fact, never in the past had there been the degree of coördination in railroad traffic that there was in the summer of 1917. Yet never before during the summer season had American railroad traffic been so congested. Food and coal were not moving as they should during this normally slack period, and the experienced traffic man looked forward to the coming winter with dismay.

His worst fears failed to picture what actually occurred. The winter of 1917-1918 descended with great severity and continued for weeks and months the coldest and stormiest in the annals of the Weather Bureau. Rail transportation faltered in December, 1917, and paralysis traveled back along its members from a dozen foci. The congestion of the previous winter had been light in comparison. The accumulation of freight at the Atlantic seaboard reached the stupendous total of 44,320 carloads, or approximately 2,000,000 tons. Tracks were so crowded with cars that it was virtually impossible to extricate properly distributed cargoes for the loading of vessels. Freight had backed up from New York to as far west as Pittsburg and Buffalo. The railroad men, trying desperately to secure cars and room, and realizing the hopelessness of waiting for the congestion to be untangled, unloaded freight by dumping it unceremoniously on the ground beside the tracks.

Meanwhile, with Europe in a desperate way for food and munitions and with our own Expeditionary Forces growing and demanding a continued and uninterrupted flow of supplies, at New York alone there were over 200 ocean-going ships anchored or tied up to the docks, unable to move because of either lack of cargoes or lack of coal for their bunkers. And this occurred at a time when the inroads of the German submarines had brought the Allies' shipping to a dangerously low mark; a time when the continuance of military operations against Germany depended upon the ability of the mariners to keep their ships moving at sea, or loading or discharging cargoes in port, every minute of every day. The immediate result of the congestion was a tremendous drop in the export cargo tonnage. When the late Lord Rhonda, the British food controller, viewed the American export figures for December, he sent to Washington a memorable message: "Unless America can increase in January the quantity of supplies sent in December, I am unwilling to guarantee that the Allied Nations can hold out." Rail transportation, because of its method of administration, had failed in the nation's greatest emergency. For America, it was the darkest hour of the war.

On December 28, 1917, the Government stepped in and took charge of the railroads.

Photo by Underwood & Underwood, N. Y.

WOOD AND CANVAS CONSTRUCTION, CAMP MILLS

Photo by International Film Service

CANTONMENT CONSTRUCTION, CAMP DEVENS

Photo by Signal Corps

LOADING PLATFORM AT ARMY DEPOT

From The War College Collection

ARMY FREIGHT LOADED AT A MUNITIONS PLANT

CHAPTER X

THE ARMY SOLVES ITS FREIGHT PROBLEM

THE War Department responded to the government seizure of the railway lines by establishing, within the General Staff's Division of Purchase, Storage, and Traffic, the Inland Traffic Service, to assume sole control of the transportation of soldiers and military supplies. Mr. H. M. Adams, an eminent traffic expert, was appointed director of the new organization. This action was taken on January 10, 1918, thirteen days after the Government began operating the railroads. Within the Inland Traffic Service one branch was created to supervise the transport of troops; but since the American Railway Association had already built up an expert and effective machine for conducting troop travel, and since the new United States Railroad Administration, the federal operating agency, had adopted this machine bodily as its troop-movement section, the Inland Traffic Service required for its passenger branch only a single officer, with his few necessary assistants, to act as the Army's supervisor of troop travel and as the official forwarder of travel orders from the Adjutant General to the troop-movement section. The rest of the extensive organization of the Inland Traffic Service devoted itself exclusively to the transportation of military freight.

The traffic situation was now in its most hopeless stage. At New York there was an accumulation of nearly 3,500 carloads of war department freight, stored in cars caught and held fast in the crush, unloaded in warehouses or upon the unsheltered ground, or stacked upon river and bay piers jammed to capacity with cargo which awaited shipment. The total amount of all government freight at the North Atlantic ports, including supplies for the European Allies, but excluding grain and coal in bulk, was approximately 30,000 carloads. The incessant blizzards of the winter had banked up the drifts and solidified the stagnation on the rails. Cars of coal came up to the docks at Norfolk frozen—a most unusual occurrence, partly because weather severe enough to freeze coal in cars is seldom experienced so far south as Norfolk, and partly because the lower Chesapeake, the best-equipped coal-loading harborage on the Atlantic coast, can handle cars so rapidly that as a rule they do not remain on the sidings long enough for their coal to freeze in any sort of weather. The export of A. E. F. supplies, which had reached nearly 178,000 tons in December, fell in January to 119,000 tons.

The task which devolved on Mr. Adams and the organization of expert traffic men which he quickly built up was to extricate the army freight from the jam, open up channels through the congestion, and provide for an expansion of traffic commensurate with the multiplying needs of the A. E. F. There was given him no time for extended study of the situation or for a leisurely promulgation of rules that should bring about the desired results. He not only had to effect complete relief, but he had to do it instantly and in the face of the worst weather conditions which eastern railroading had known. Severe cold appreciably slows down transportation, even with other factors normal. Switches freeze and become hard to operate, brakemen and trackmen are hampered in their movements by heavy clothing, and low temperatures decrease the pulling power of a locomotive. Add to these conditions snowdrifts so deep that passenger trains are stalled in the streets of Syracuse, New York, and you have some idea of the difficulty of the Army's task.

There was, however, this element of hope in the situation: every traffic man knew what was the trouble and why the jam had occurred. The railroads themselves, being shackled by the anti-trust laws which prevented combination for the sake of efficient operation, had been unable to apply the remedy. The Government's seizure of the lines had for the time being wiped off the statute books the laws which prevented unified operation; and the Inland Traffic Service, through the medium of the Railroad Administration, could now administer the remedy at will.

The trouble was primarily the lack of coördination between shippers and consignees. Roads which originated traffic naturally put upon their rails as much business as they could secure, whenever they could secure it, regardless of conditions at the terminals into which that business had to flow. Many of the war contracts provided that the Government should pay for supplies as soon as they were loaded on freight cars at the factories. Holders of such contracts were in a hurry to load, whether or not the Army was yet ready to use the goods. The various production bureaus felt the spur of competition to make a creditable showing by putting their freight on the rails as quickly as possible, although often there was no urgent demand.

The Government, through the War Industries Board, had established priorities in the rail shipment of materials to the munitions plants; but up to this time there had been no single official organization which determined priorities in the overseas shipment of finished supplies. Each producing bureau had rushed to the ports as much of its *matériel* as it could secure from the factories, regardless of the place held by the goods in the whole military scheme.

To control car congestion, the railroads themselves had possessed only the clumsy remedy of the general embargo. They could clamp down this lid on a congested district, dam back all traffic until the congestion inside the embargo had been relieved, and repeat the process whenever necessary. Embargoes of greater or less severity had been imposed around the Atlantic ports for several months before the end of 1917; and when the congestion of the early winter became acute, these embargoes, particularly at the port of New York, were made, so far as the commercial shipper was concerned, complete. The principal exception was all war department freight.

Before the time of government operation, when the Committee of Five was still attempting to handle the situation, an army production officer had only to go before the railroad authorities and make representation that the products in his charge were an immediate necessity for the A. E. F., and he could secure orders permitting his shipment to move into the congested tidewater district. Each production officer was naturally an enthusiast for his own specific responsibility and prone to exaggerate, honestly enough, the relative importance of his own supplies. The railroads and, later on, the United States Railroad Administration had no means of judging whether the production bureau's arguments were sound. They could only grant the order which opened the embargo for the proposed shipment. At tidewater the embarkation officers formed their own opinions as to the urgency of export. They might put certain incoming shipments aboard transports at once, or they might leave them on the tracks while more necessary freight appropriated the narrow passages through the congestion.

The organization of the Inland Traffic Service gave the War Department, for the first time, a central agency for regulating the rail movement of export supplies according to their importance in the maintenanee of the A. E. F. The secret of orderly traffic was complete control throughout transit from beginning to end—control at the source exercised from the terminal. But before such control could be exerted systematically it was necessary to formulate regulations; and, pending the day when the Inland Traffic Service could promulgate its system, Mr. Adams asked the Railroad Administration to make complete against all war department traffic the embargo around the ports. Exceptions to the embargo were to be made only upon request of the Inland Traffie Service itself.

Thereafter, when a producing bureau asked for railway priority to a port, it had to make a case for itself before the Inland Traffic Service, which was in close touch with the whole military situation and competent to determine the urgency of any request. The Service was also in immediate liaison with the Embarkation Service, which had charge of the ocean terminals and of the army transports. Thus it was advised at all times whether commodities, in case they were passed through the embargo, would find ship room.

By these means the new Service held back the flow of freight until it could secure elbow room at the ports. Its earliest actions were specific treatments of special cases as they arose. It hastily analyzed the freight already at the ports, whether in cars or unloaded. The more urgent freight was rescued from the jam and placed upon the docks and piers. The less urgent was unloaded and stored. It must not be supposed that the whole congestion was cleared up immediately. But these emergency measures continued only until a semblance of order had been secured and more room obtained in the railroad yards. Then the Inland Traffic Service was ready to put its permanent system into force.

On the 18th of February, 1918, the Inland Traffic Service issued its first two orders, which set forth the rules under which war department freight would thenceforth move on the rails. These rules were so well considered in advance that it was never necessary to change them. They became the permanent law which brought order out of chaos. Later amplifications of them extended the system throughout the United States.

The first of these orders, No. 1, was a notice to the bureaus of the military organization that the Inland Traffic Service had assumed full control of all military traffic, both freight and passenger. All army negotiations with the railroads and other private transportation companies were to be conducted solely by the Inland Traffic Service. The Army had made a practice of shipping freight to terminals without specific billing directions as to the delivery of such freight, expecting to supply the delivery instructions later, after determining where the property could be stored or received for overseas shipment. This practice, because it had resulted in holding loaded cars in terminal yards, was an aggravating element in the car congestion. Order No. 1 forbade its continuance. The order also prohibited the shipment of freight to any destination for diversion or reconsignment, another practice which had sprung up in military traffic. The order forbade the shipment of property to central points for inspection and reshipment, by requiring inspection prior to the original loading upon freight cars. The order shut down on the extravagant use of the railway express in war department traffic. It set forth other rules, all of them directed to the relief of rail congestion.

On the same day, February 18, the Service issued its Order No. 2, the actual operation order. This established a permanent embargo against war department freight at New York and its environs, Philadelphia, Baltimore, the whole port region of lower Chesapeake Bay, and other principal Atlantic ports. From this embargo the order excepted only construction material consigned to the various military building projects within these regions.

After it had fastened down the embargo—and this embargo was to continue until the war was over and demobilization came to an end, extending eventually to the principal producing and receiving centers of the entire United States—Order No. 2 then went on to outline the procedure whereby army freight might pass through. The open sesame was the War Department Transportation Order, a new thing in railroading, and later known in traffic by its initials, W. D. T. O. Government contractors and producing bureaus were forbidden to place a pound of freight on any car billed for the proscribed zones unless they possessed Transportation Orders issued by the Inland Traffic Service in Washington and signed by an authorized officer in the Service.

One could not evade the system of control. A shipper might manage to get hold of a car, but he could not load it with war department freight and expect to dispatch it into one of the embargoed areas, because the railroads themselves, by order of the Director General in Washington, were forbidden to receive freight billed through the embargo, unless the shipper produced a War Department Transportation Order. The regulations prescribed that each bill of lading for such freight must bear upon its face the serial number given by the Inland Traffic Service to the Transportation Order permitting the shipment to be made, and no bill of lading was legal without this number.

In order to secure a War Department Transportation Order the shipper, who might be a government bureau, a field officer of a producing bureau, or a private contractor, had to make application to the Inland Traffic Service in Washington. In this application the shipper was required to describe the property to be sent, to give its weight in tons, and, if it were destined overseas, to supply its cubic measurements for the convenience of the Department's ship loaders. The shipper had to describe in his request the sort of cars he needed, estimate the period within which the entire consignment would leave his freight platforms, and name the consignee and the railroad or other carrier which would probably handle the shipment at the delivery point. He also had to certify that the property would be accepted by the consignee upon delivery and promptly unloaded from the cars.

The Inland Traffic Service studied these allegations in the light of its sources of information and decided whether there were need for the shipment at that particular time. Then it went further. It referred the request to the embarkation service officers at the ports, who determined whether there would be ship room or storage room for the consignment. The Embarkation Service, if it approved the shipment, issued a so-called release, upon which the Inland Traffic Service based a Transportation Order, which it forwarded to the shipper.

The War Department Transportation Order, addressed to the railroad which served the shipper's plant, requested the delivery to the shipper of a sufficient number of cars to handle the consignment. This order was mandatory upon the railroad. The shipper was instructed in the order itself to display upon the face of all bills of lading and way bills the serial number of his order, and the railroad was forbidden to accept the shipment upon a bill which did not display the number. The shipper presented the order to the local agent of the carrier. If he experienced difficulty in securing equipment, he was instructed to report the fact to the Inland Traffic Service in

Washington, which thereupon took up the matter directly with the Railroad Administration. The shipper was forbidden to use cars supplied upon a Transportation Order for any other purpose than that named in the order. The order allowed the shipper a designated period—usually fifteen days for export freight and thirty days for domestic—in which to load the consignment upon cars. The order became automatically void within forty-eight hours after the expiration of this period. If for any reason—whether through the fault of the shipper in failing to get the freight ready or the failure of the railroad to provide the equipment—the consignment were not loaded and accepted by the carrier within the time specified, it could not be moved until the shipper had secured a new order.

This system of port releases and rail Transportation Orders quickly brought relief to the military freight congestion at the seaboard. It provided at once for the flow of materials into the ports in a regular order prescribed solely by the necessities of the A. E. F., the ability of the port loading forces to handle the commodities, and the capacity of the ocean tonnage available. Fresh military freight no longer accumulated in the yards; the railroad forces at the terminals were able to make heavy inroads upon the congestion already existing; and, although the severe weather conditions persisted, the traffic steadily cleared up. Each day brought improvement. By March 30, 1918, the accumulation of nearly 3,500 carloads of war department freight at the northern Atlantic ports had been reduced to less than 2,000, and all the while the transportation authorities had been handling an expanding stream of exports through the embargo. Expressed in other terms, stagnant war department freight at the ports when the Inland Traffic Service took hold amounted to some 150,000 tons, or a greater weight than the total freight shipped to the A. E. F. in the entire month of January. Of this property about 50,000 tons, or one-third, had been unloaded on the ground in order to release cars and provide track room in the yards. By the end of March this accumulation had been reduced at least half.

Meanwhile the Railroad Administration was applying similar measures of relief to all other export freight, and the congestions at the seaboard rapidly cleared away. The accumulation of more than 44,000 cars at the ports in December was cut down to little more than 30,000 by the first of April. At New York the congestion had involved more than 26,000 cars. By April this accumulation had been reduced to 16,000, and by the following October it was below 10,000—a normal number, considering the volume of traffic.

The relief afforded to conditions at the ports may be measured in the export figures of supplies shipped to the A. E. F.* December had been the heaviest export month for the War Department, with approximately 178,000 tons of army supplies shipped overseas. In January the export fell, as was stated above, to 119,000 tons. In February the figure was over 233,000 tons, or almost twice as much as the ports had been able to handle in January. This is the best testimonial to the immediate relief which the Inland Traffic Service was able to supply. These figures are but cold mechanical expression of an achievement co-equal in merit with any other incident of the dispatch to Europe, in little more than a year, of 2,000,000 American troops and 6,000,000 tons of army supplies. In March, with weather conditions still adverse, the export of war department freight approached the 300,000-ton mark, and in April it surpassed this record by over 80,000 tons. With exports continually expanding in volume, car congestion disappeared, and order took the place of confusion; until, in November, the War Department sent over 800,000 tons of freight to the A. E. F. All other essential exporting, including food and munitions for the Allies, kept pace with this growth. For such an outcome, one thing and one only was to be thanked: complete control of inland traffic from shipping point to destination.

At first the control of army freight was exerted only over exports. But as the Inland Traffic Service got its hand in and by hard work straightened out the traffic at the ports, it found

* See Appendix C.

time to take up other matters; and then it rapidly extended the control system by a series of orders expanding Order No. 2, until, eventually, embargoes surrounded every principal receiving point for military freight in the United States. It became almost impossible to ship army supplies anywhere without running into an embargo and the necessity for securing War Department Transportation Orders. The Inland Traffic Service threw its embargoes around the southern ports and those of the Pacific coast, around the principal training camps, around interior quartermaster depots, around the larger artillery proving grounds, and around some of the greater munitions plants—in short, wherever military supplies converged and were likely to produce car congestions.

Simultaneously it expanded its own organization. During its first few weeks of existence the Inland Traffic Service made use of the traffic organization of the Ordnance Department. Then it began stationing its own officers, commissioned or civilian, at the principal receiving and dispatching centers in the army supply system, until eventually it maintained an office at every principal freight-handling point. This expansion began in earnest on May 1, 1918, on which day the service established offices at New York, Chicago, and Jacksonville, Florida, under the direction of experienced traffic men. By August 1 the service had set up branches at all the principal industrial cities of the country, as well as at all the ocean terminals. The outside organization handled the details of issuing Transportation Orders, kept the central office advised of traffic conditions in the several districts, and in general facilitated the smooth and rapid operation of the system.

It should be understood that the control system applied only to the transportation of military freight in carload lots. The Service at no time attempted to direct the transportation of consignments that did not fill up at least one freight car. But, as a fact, the great bulk of military supplies moved in solid cars. Even in the aggregate, the less-than-carload shipments of army materials were negligible; and they had never been a chief contributing cause of freight congestion. The reader will see here an analogy to the military handling of passenger traffic: the troop-movement office, it will be remembered, did not concern itself with the travel of soldiers in parties of less than fifty men.

As an aid to the management of military freight traffic, the Inland Traffic Service created within itself the useful car-record office and tracing bureau. This subsidiary kept track of all loaded cars of army freight and watched their progress from origin to destination. The work of the car-record office required the energies of a force of from fifty to seventy-five men, most of them men of railroad experience. The bureau worked out a method which enabled it to know at all times the approximate location of every war department car. Its field organization amounted to a live-tracer system stationed permanently along tracks at the principal junctions. Each agent of the office reported by telegraph or long-distance telephone upon every important shipment when it reached or left his station, and thus the central office at Washington usually knew within, at most, 200 miles the whereabouts of every one of its cars. With such a system, it was impossible for cars to get lost in transit or even to be greatly delayed.

The car-tracing bureau was of great service to the Government in that it enabled the quick diversion of freight, if that at any time became necessary. And it was frequently necessary. The presence of enemy submarines off the Atlantic coast during the summer of 1918 and the continued inroads of the U-boats upon allied shipping on the other side of the Atlantic often entirely changed the sailing plans on this side, in which event it became necessary to divert supplies from one port to another. With immediate knowledge of the approximate position of all freight on the rails, the Inland Traffic Service could direct the flow of traffic into one port or another, exactly as a fireman aims the nozzle of a hose. Thousands of cars were, in fact, so diverted. At one time, cargo transports were approaching the Atlantic coast to make up a convoy which was to load at New York with supplies for the A. E. F., including a large quantity of flour; and a hundred or more cars of this export flour were on the rails traveling from Minneapolis to the port. At the last moment Washington issued an order which required the flour ship to load at Newport News. Under the old system, this change would have resulted in great delay while tracers were running down the carloads of flour and changing their destination. Many of the cars, doubtless, would have reached New York and lost themselves in the accumulation of freight cars in the yards. As it was, the Inland Traffic Service caught all of these hundred cars within a few hours and diverted them so that they arrived at Newport News almost as quickly as they would have arrived at New York, allowing only for the greater distance they had to travel.

Baled hay was a commodity frequently subject to diversion, simply because the number of horses and mules at a given camp was continually fluctuating. A camp quartermaster might estimate the amount of hay which the camp animals would require for a certain period and make requisition for this forage. By the time the commodity had been shipped, the exigencies of the service might have transferred many of the animals to some other camp, and the quantity of hay *en route* would no longer be needed at that point. Then it became the duty of the car-record office to find the shipment and divert it to some point of fresh demand. The sooner such diversion was accomplished, the sooner the railroad equipment would be released for other work. A great deal of the efficiency of military control of freight traffic was due to the fact that the Inland Traffic Service was in such instant touch with the traffic on the rails.

The survey made by the Service immediately after its establishment in January, 1918, uncovered a number of traffic abuses which had sprung up under the cloak of military necessity. It is hard for the average citizen to understand the spirit of a person who can see in the extremity of a country fighting for its existence an opportunity to make money by fraud. Yet such scurrilous practice was by no means rare. It existed, not only in transportation, but elsewhere in war industry. The munitions production bureaus soon learned that there were manufacturers base enough to secure government contracts solely for the sake of the right which such contracts gave them to draw upon the supplies of raw materials. Such a manufacturer, once he had obtained his raw materials, would default in the contract and use the materials in making up finished products for his private customers.

Fortunately, such instances were not many. But when it came to securing transportation privileges, it seemed that any number of shippers were willing to employ devious means to thwart the traffic-regulating orders. The commonest practice was to take advantage of the preference given to freight billed through to officers of the Army. This abuse occurred in the days before the Inland Traffic Service had extended its transportation order system to cover all army freight, regardless of its destination. In those times the unscrupulous shipper who had a consignment of goods for a private customer in some city would bill the freight to an army officer or war department bureau at that destination. The railroad, accepting the freight as *bona fide* military supplies, would give it preference over commercial freight and put it through rapidly. The actual consignee at the other end, who was likely the local agent of the shipper, would present the bill of lading and secure the goods. In other instances, shippers who had held government contracts continued to use the contract numbers to secure rail priorities long after they had completed their deliveries to the War Department.

In fairness to American industry it must be said that no large or reputable concern stooped to such methods. Yet the Inland Traffic Service found hundreds of such frauds perpetrated by the smaller and less scrupulous companies. Often the officers named as consignees were wholly fictitious, and the railroads themselves sometimes became suspicious of certain shipments. After the Inland Traffic Service was formed, the railroads began to report suspected shipments. The Service issued warnings to shippers throughout the country, unleashed the legal forces of the Government upon those who disregarded the warnings, and thus gradually eliminated the abuses. The extension of the transportation order system over the United States also aided in checking fraud.

Another conspicuous abuse, already cited, was the extravagant and ill-advised employment of the express service in the transportation of war department property. This abuse, although not venal, was a source of considerable needless expense to the Government. It was due, oftener than to anything else, to lack of good judgment on the part of army officers. Officers soon discovered that they could put their shipments through to destination much more quickly by express than by freight, particularly during the months when the freight traffic was staggering. There ensued the most absurd uses of the express. The carload express shipment became common; it was not unusual for an officer to ship an entire trainload of freight by express. The charges on such shipments were, of course, enormous. More important, the express system became so overloaded that it could not perform its really useful function of putting through emergency shipments with dispatch. As the indefensible practice continued, it actually became true that express service was often slower than freight.

Officers who abused the express privilege seemed on occasion to exercise no judgment or discrimination at all. Numerous heavy articles for the A. E. F., such as automobiles, machinery of various sorts, freight cars, and box cars for the American military railroads in France, were sometimes shipped to the ports as parts, there to be set up and assembled before being loaded on the ships. Zealous supply officers frequently shipped component parts of these articles by express before some of the more important parts had even been manufactured. One officer attempted to ship to Brooklyn by express a consignment of electric warehouse trucks for use at the new army base in South Brooklyn. At that time the warehouses in Brooklyn were nothing but concrete shells, empty and even floorless; and the working force at the base was to have no use for warehouse trucks for another half year.

The prize example was the shipment by an American officer of nearly fifty express carloads of iron cots from Pittsburg and Atlanta to a hospital under construction in Texas. The hospital was not only not ready for the beds, but there was not even unloading space for them, and they had to remain in the express cars at El Paso for six weeks before covered storage could be provided.

The express abuse was broken up by an order of the Inland Traffic Service, an outright prohibition of the shipment by express of any war department consignment weighing more than 1,000 pounds. To send any heavier consignment by express, it was necessary for the shipper to secure a War Department Transportation Order, which gave the Inland Traffic Service the right of visé in the matter. This measure cleared up the express situation at once, saved enormous sums of money for the Government, and enabled the War Department to make full use of the express as an emergency service.

After the Inland Traffic Service had extended its organization throughout the country, it became a master of American freight traffic, although war department traffic *per se* was but a minor percentage of the total volume. The Army gained this eminence largely because of the personality of the chief of the Service, Mr. H. M. Adams. Mr. Adams was universally recognized as one of the ablest traffic men in the United States. He had been the vice-president in charge of traffic of several important railroads, and after his retirement from the government service in 1919 he became vice-president and general traffic manager of the Missouri Pacific Railroad. Railroad men not only had complete confidence in his judgment, but they recognized in him a man actuated only by the highest of patriotic ideals and by a stern sense of duty.

The relation of the Inland Traffic Service to the railroads was technically advisory and not mandatory. The mandatory power existed only in the United States Railroad Administration. The Inland Traffic Service could order only the actions of war department shippers and consignees: the coöperation of the railroads it had to request through the Railroad Administration. The Railroad Administration controlled the war traffic through a committee of traffic men, each representing one function of the Government. Mr. Adams was the war department representative on this committee, to which the Navy, the Food and Fuel Administrations, the War Industries Board, and all the other war organizations which had a special interest in transportation sent their traffic officers. These men were the best the nation's railroads could supply to the Government. It is no disparagement of them to say that Mr. Adams was at least the peer in ability of any man on the committee. Because of his commanding position in the railroad world, his advice or recommendation acquired the weight of a command.

War department traffic was so interknit with other traffic that it was impossible to separate it from the operational whole. The army traffic could not be directed effieiently unless all the other traffic were efficient and minded to coöperate. At one time the War Department had great difficulty in securing lumber for its construction projeets; in fact, there was no lumber to be had. The producers preferred to utilize the limited transportation facilities to take care of their own customers ahead of the Government. Mr. Adams employed drastic means to correct this situation. He asked the Railroad Administration to put down a complete embargo upon the shipment of lumber to any consignee other than the Government. The Railroad Administration acceded to this request and maintained the lumber embargo for about five months. This was the greatest commodity embargo imposed during the war through the agency of the Inland Traffic Service; but it was only one of several for which Mr. Adams was responsible. When, at the expiration of his service, the Government awarded a Distinguished Service Medal to Mr. Adams, it was the consensus of opinion in the Transportation Service that honor had been rendered where honor was due.

Prior to the assumption by the Government of the operation of American railroads, every shipper in the United States had possessed the right to route his shipments to their destinations exactly as he chose. The routing of freight traffic is a distinct branch of modern industrial science. All large concerns maintain their own traffic experts, a considerable part of whose duty it is to see that freight is routed for the quickest possible journey, or for the lowest possible rate, or for both. It was this shippers' privilege, combined with the initiative of the independent railroad lines in securing business for themselves, that resulted in the periodical congestions of certain roads. The United States Railroad Administration at once ended this routing privilege and allowed the traffic to follow lines of least resistance—that is, to flow over the lightly used tracks when the more heavily traveled ones were taxed to capacity. In conformity with this new practice, the Inland Traffic Service established its own routing section, which acted for war department freight much as the routing division of the troop-movement office did for the human personnel. The care of the freight-routing section was directed principally to the ends that terminal lines should not be congested and that service generally should be improved. Another and more specific function of this service was to operate special freight trains carrying high explosives or poisonous gas—things forbidden space in mixed trains—and also to operate special trains loaded with supplies peculiarly important to the Army.

Just as the troop-movement general agents at the camps procured passenger equipment and supervised the loading of men on trains, so the traffic agents stationed at these same camps by the Inland Traffic Service became the local freight managers. They secured the needed cars, arranged for the switching, took charge of making out freight bills and of accomplishing bills of lading on inbound shipments, and saw to the prompt settlement of freight charges. No figures yet made available show the total war department freight traffic in the interior of the country; but Appendix D of this work, which reproduces the traffic statement from Camp Grant, at Rockford, Illinois,—a cantonment which may be considered typical of the training camps,—indicates the amount of work which devolved upon the Inland Traffic Service's manager at a single camp.

When the export movement was at its height, in 1918, the

average daily volume of war freight of all kinds going into New York was 1,000 cars. This was at least the equivalent of 20 freight trains daily. As many as 6,500 carloads were within the New York yardage at one time, but in orderly movement, so that they by no means constituted a congestion. In order to keep track of the freight situation, the Inland Traffic Service maintained at New York a large force of civilian and commissioned personnel to check up the traffic continually and keep Washington informed. Similar, though smaller, organizations at the other principal ports rendered direct reports twice a day by telegraph or long-distance telephone.

The handling of traffic in the interior portions of the country was in itself a vast undertaking. The war set in motion traffic currents where none had run before. The establishment of such enormous institutions as the powder plants at Nitro, West Virginia, and Nashville, Tennessee, and the projected ordnance plant at Neville Island, near Pittsburg, suddenly created new production areas; and every training camp in the country was in effect a new city consuming great volumes of supplies. The powder plant at Nashville became an industrial center as large as Wilmington, Delaware, and required the construction of nearly two hundred miles of switches and sidings in its railroad yards. Some of these new producing and consuming centers were situated at places without adequate railroad facilities. Camp Shelby, Mississippi, for instance, was located on a logging railroad which had operated only one commercial train a week before the camp was set up. It was necessary for the Inland Traffic Service to provide the facilities to handle the freight traffic at such points. This it sometimes did by operating shuttle trains in and out of the camps, to bring in loaded freight cars from the nearest points which had good railroad advantages.

There arose, too, a tremendous import traffic in raw materials for war department plants. Where America had imported a hundred tons of Chilean nitrates in the past, she now began importing four hundred tons. The nitrate shipments came in mostly through southern ports. When the system of control was in full operation, it required a Transportation Order to move a shipment of nitrates out of a port to an interior powder factory. The system worked both ways: it prevented congestion at the interior producing point just as efficaciously as at the port.

CHAPTER XI

SOLVING THE NATIONAL WAR FREIGHT PROBLEM

HOWEVER admirable the system of the Inland Traffic Service for the control of war department freight traffic, it would have had no such decisive results unless the general freight congestion in the United States had been relieved. While the War Department was putting into effect its own regulations for the efficient transportation of military property, the United States Railroad Administration was applying similar measures to the national traffic situation.

War department property constituted probably not more than one-tenth of the volume of freight involved in the congestion at the Atlantic ports, and perhaps even a smaller percentage of the interior accumulations. The War Department insisted upon preferential treatment for its cars in transit; but there were other essential governmental agencies, all of them similarly pressing for the movement of their various commodities. The Navy had to move large volumes of supplies, not only for its ships, but also for its training camps and other land establishments, which multiplied after the declaration of war. The Shipping Board was encouraging, financially and by direct coöperation, the construction of shipyards up and down both coasts and the building of ships in those yards. Shipyard construction materials and ship timbers, ship steel, and ship machinery made up an impressive part of the total volume of traffic. The Food Administration, charged as it was with the duty of supplying food to the Allies, pressed for the transportation of its commodities. The Fuel Administration was no less urgent that the mines should have full traffic facilities. The War Industries Board was concerned with the freight movement of materials to the munitions plants; and so on. Each of these great organizations maintained an aggressive traffic bureau and fought against the others for priorities in transit. The competition of these agencies contributed largely to the throttling of traffic at the onset of winter in December, 1917. The railroads and their emergency war committees had been unable to decide equitably among such competitors; moreover, they had possessed no mechanism for the control of traffic at its sources. They could not forbid shipment: they could only stop traffic when it reached the barriers of an embargo.

One of the first acts of the United States Railroad Administration, therefore, was to coördinate the entire government demand for freight-movement facilities. The Director General appointed a manager of inland traffic for each principal war agency. Mr. Adams was the appointee for the War Department; the others were men of like caliber. The traffic manager appointed for the Navy was Mr. H. P. Anewalt, who had been general freight agent of the Santa Fé Railroad. The Director General took from the Burlington Railroad its vice-president, Mr. C. E. Spens, and made him traffic manager for the Food Administration. Mr. F. M. Whitaker, vice-president of the Chesapeake & Ohio, became traffic manager for the Fuel Administration. Mr. T. C. Powell, vice-president of the Southern Railroad, served the War Industries Board. Mr. D. L. Gray was taken from the Erie Railroad, where he had been the general traffic manager, and put in charge of the traffic of the United States Shipping Board.

These men were among the greatest transportation experts of the country. They and their staffs came to the Government as loans from the railroads. They were put on the pay roll of the Railroad Administration, but each one of them was subject to the orders of the war organization in which he was placed. Meeting once a week, these managers constituted, within the Railroad Administration, a traffic committee for arranging priorities and allocating the transportation facilities among the several bureaus and departments. This was probably the most expert traffic committee ever assembled for executive action. Every man on it was preëminent in his field.

The weekly committee meeting knit into one administrative unit the entire sum of the Government's traffic interests. Although in general the War Department was given priority, particularly in shipments for export, the committee could and did drive in any particular direction where energetic action was needed. Its flexibility was never better shown than in a traffic accomplishment late in the winter of 1918, when the United States literally saved the Allies from collapse by an extraordinary delivery of food at a crisis.

In February, 1918, the ambassadors of Great Britain, France, and Italy jointly sent an invitation to the Director General of Railroads to hear a message of grave importance sent from their respective prime ministers. Mr. McAdoo, unable to attend the meeting, sent Mr. C. R. Gray, the Director of Railroad Operations. To Mr. Gray the ambassadors read messages to the effect that America had fallen nearly a million tons short of the program of supplying foodstuffs to the Allies. The notes were strikingly similar in their expression of hope that in some way America might be able to make good the deficit. One dispatch concluded with the statement that the dearth of wheat abroad was "the greatest danger threatening the allied nations of Europe." Another prime minister cabled that, if the deficit could not be made up, the war would be over by April 1. There was just one month in which to save the situation.

Through the traffic committee the Food Administration readily secured priorities to cover this enormous shipment. The various federal boards, departments, and administrations represented on the committee agreed to use other lines for the transport of their Atlantic seaboard freight, leaving the New York Central Railroad artery, with all four of its tracks if necessary, open exclusively for the transportation of export wheat and other foods. In the emergency the Railroad Administration committed the traffic heresy of sending box cars empty from the East to the grain centers of the Middle West. It also adopted the drastic rule that until further notice there should be no commercial loading of box cars anywhere in eastern territory, except for the movement of essential food and fuel. The cars thus made available were rushed westward in solid trains; and, although the blizzard conditions still persisted, the New York Central was kept open, and a flood of wheat descended upon New York.

The British, upon whom devolved the burden of freighting this food across the ocean, were caught unprepared for what actually occurred. The Admiralty had sent to New York cargo vessels of total capacity much beyond what it thought the Americans could fill. Yet by March 15, one month and one week after the cry for help had been heard, every available ship had been loaded and sent out, every grain elevator on the Atlantic coast was crammed to capacity, and in addition the excess supply of food at the seaboard filled 6,615 cars on the sidetracks. In fact, it became necessary for the Railroad Administration to threaten an embargo upon export food unless the Allies could improve the ocean shipping situation. In the three or four weeks of greatest activity, civilian traffic was suspended almost altogether; indeed, there was a period of more than a fortnight when, except for the transportation of necessary food and fuel, print paper, and freight of a few minor classes, the ordinary commerce of the East and the Northwest was totally arrested. Of government property, only that of the War Department moved in this interval. Such a feat would have been utterly impossible in the days before the united control of freight traffic.

The Railroad Administration adopted for all inland commerce a system almost identical with that by which the Inland Traffic Service handled war department freight. The Transportation Order, to be sure, did not exist in commercial traffic, but in its place appeared the shipping permit, which was essentially the same thing. The Railroad Administration maintained, first at New York and then elsewhere, as the system expanded, local freight committees of railroad representatives.

Before a producer in any part of the country could ship freight into a terminal, or even secure cars from the railroads, he had to obtain a permit from the freight committee at the terminal. This committee, in intimate touch with its local situation, knew approximately whether freight could be handled promptly upon arrival. The shipper who applied for a permit was required to show that his freight would be accepted and unloaded immediately by the consignee. The committee could check up his statements. The general system of embargo was extended to cover all centers where congestions were likely to arise, and the permit was the only key to the door.

When the Railroad Administration took hold, nearly a quarter of a million freight cars were caught in the various congestions and accumulations at the seaboard and throughout the United States. That is to say, there was that number of cars loaded and on the rails, but not running currently with the traffic, and serving perforce as mere places of storage for the goods which they contained. This was ten per cent or more of the entire car equipment of the United States. The removal of so many cars from the traffic constituted a grave embarrassment to the whole transportation system. The centralized control of the railroads reduced this accumulation steadily, until in the following winter (1918-1919) there were less than 30,000 cars so tied up.

The management of the rolling stock was in the hands of the car-service section of the Railroad Administration, an organization which had formerly been the commission on car service, a subsidiary branch of the American Railroad Association's Committee of Five. This organization herded all freight cars together according to class, without regard to ownership, and made the entire American freight equipment fluid for the first time in history, sending rolling stock at different seasons and intervals to sections of the country where large freight movements were impending. The result was that, in the summer of 1918, in spite of the enormous war traffic, the American public, for the first time in years, heard nothing about car shortages or the inability of the railroads to move the seasonal tonnage.

Wheat, for instance, had for years clogged the railroad channels at the time of its heaviest movement, creating car shortages outside the wheat belt and congestions within it that hampered the easy flow of the crop to the market. In September, 1918, when the wheat crop started to move, the Railroad Administration threw embargoes around each of the twenty largest receiving wheat centers of the United States. Traffic committees were established in these centers, and the shippers of grain were required to secure permits from the committees in order to obtain cars. This system kept the terminals clear. Meanwhile the car-service section rounded up box cars for the grain districts from wherever they could be got, and the wheat crop was moved with unprecedented speed and efficiency. Formerly, just in advance of harvest time, the railroads had seized whatever empty cars were upon their rails and held them for the protection of their own shippers during the crop movement. Under federal control, empty cars moving westward for grain could not be stopped in Ohio or Illinois until the grain in those states was actually at the rails for loading. This prohibition permitted plenty of cars to go on into districts where the harvest had already begun, and federal control followed the harvest northward with equipment as it progressed.

The car-service section maintained an officer in Chicago who had charge of the operation of the common pool of refrigerator cars. Formerly each road which owned refrigerator cars protected them from use by other roads. The pooling of rail refrigeration put adequate equipment at the disposal of any given shipping territory; and in 1918, almost for the first time, the perishable food crops moved expeditiously.

The management of the railroads as a unit resulted in other great reforms in freight traffic. Summer is ordinarily the slack time on the railroads, and the Railroad Administration took advantage of this fact to anticipate the forthcoming industrial demands for raw materials. The paper mills in the northern part of the United States import great quantities of pulp wood from Canada. Ordinarily this traffic has been heaviest in early autumn. By May, 1918, the car-service section began shipping empty box cars to Canada for pulp wood. The movement continued all summer, involving the transit of over 70,000 cars, an average of over 500 cars, or ten trains, a day. By October 1, 1918, the American supply of pulp wood in Canada was entirely cleaned up for the first time in many years, and that in the midst of an abnormal war traffic which might have been expected to impede ordinary commercial transportation.

The Railroad Administration induced the American chemical plants, which had suffered severely for raw materials during the winter of 1917-1918, to lay in their phosphate rock from Tennessee and Florida during the summer months. In October, 1918, there was a reserve of half a million tons of rock at the chemical plants, and meanwhile the daily shipments were meeting current requirements.

The war vastly increased the demand of American chemical plants for sulphur, which is produced in Louisiana and Texas. The munitions-manufacturing program of 1918-1919 called for the shipment of 1,800,000 tons of sulphur from the Gulf coast to the mills in the North. The Shipping Board agreed to transport 600,000 tons by water, leaving 1,200,000 tons for the rails. This meant an average haulage of 100 carloads a day—four times the rate at which rail transportation had been required to bring sulphur north during any pre-war year. By meeting the situation in advance and by directing the pooled car equipment into the sulphur-producing region, the Government's program had been completely met by October 15; and the railroads even handled more than they had contracted to do, for, as it turned out, the Shipping Board was unable to carry its full allotment.

The car-service section also watched new producing areas created by the war effort and kept them well stocked with rolling equipment. One of the greatest of these areas was the fir and spruce district of the Pacific Northwest, from which great quantities of ship timbers, construction lumber, and the vitally necessary aviation lumber were sent to various consuming districts. In all, 150,000 cars of lumber moved out of this district by the middle of November, 1918; of these, 12,700 carried spruce to the airplane factories. The average daily movement amounted to 813 cars.

Coal was another commodity which required most careful control in transit. Competition in the bituminous coal industry had set up a great deal of cross-hauling, a sheer waste of transportation not to be tolerated in war times. Illinois mines would sell coal to the East, Pennsylvania mines to the Northwest, and so on. The Railroad Administration, in coöperation with the Fuel Administration, mapped the country into zones and forbade the shipment of coal outside these zones. Numerous companies and railroads which handled coal to the lower Lake and Atlantic ports owned their own coal-car equipment and, in the days of individual operation, clung to it jealously. Traffic inefficiency resulted; for one owner of cars might have more business than he could attend to, while another would not be operating to capacity. All these cars were now pooled and operated as a common stock under the direction of the car-service section.

Another reform of federal operation was the adoption of the so-called "sailing-day plan" for handling less-than-carload freight. Why this measure was called the sailing-day plan is an inscrutable mystery, for it certainly had nothing to do with the sailing of ships. Less-than-carload freight is profitable to a railroad because of the heavy tariffs charged for its transportation. Every railroad attempted to make its line attractive to the occasional small shipper. This it did by operating daily out of each important center at least one freight car for the reception of less-than-carload shipments. Such shipments in the aggregate seldom loaded the cars full; oftener the l.c.l. car traveled half empty. The Railroad Administration's sailing-day plan can best be described by the illustration of what occurred at such a city as Chicago. Each principal railroad operating between Chicago and New York sent eastward daily a car for l.c.l. shipments. The Railroad Administration consolidated this effort. All l.c.l. freight was shipped each day by a single railroad. On Monday, say, the New York Central would handle the cars, on Tuesday they would go *via* the Pennsylvania, on Wednesday by the Baltimore & Ohio, and so on, so that each road received its share of the haulage. This consolidation enabled the cars to be loaded full, and thus released all the excess equipment formerly required. This item may seem insignificant; but it is estimated that the sailing-day plan, in full force throughout the United States, is capable of saving the daily handling of thirty tons of freight at each of 400 points. During the war period the plan was adopted only between the principal eastern terminals.

The Railroad Administration also did for the entire government traffic what the car-tracing bureau of the Inland Traffic Service performed for war department freight: it kept track of the movement of cars and traced lost cars for all the various war agencies. This service was rendered by the car-service section, which in the first year of operation recorded the movements of an average of 2,700 cars daily, or over 1,000,000 in the entire year, a work requiring an organization of nearly 150 employees. It traced down and located about 16,000 lost cars loaded with government freight.

To the War Department itself, the car-service section rendered the special service of maintaining a steady supply of freight car equipment to the camps and other military establishments. Up to the signing of the armistice, the Railroad Administration moved to the camps, aviation fields, and other army centers over 400,000 cars loaded with construction materials, supplies for men and animals, and other commodities necessary to the maintenance of these plants.

The successful prosecution of such a war as ours against Germany required the complete coördination of all the resources of the country. One government service might proclaim in its propaganda, "Food will win the war!" Another might blazon the motto, "Coal will win the war!" or "Ships will win the war!" As a fact, no one of these elements could win the war single-handed; it could only do its vital and indispensable share. When Marshal Joffre declared that the railroads of France won the battle of the Marne, he spoke with pardonable rhetorical exaggeration. The French railroads could not have won the battle unless there had been trained troops for them to transport and food for those troops to eat and ammunition for their guns. The railroads of the United States could not win the war; but, like any other indispensable agency, they could easily enough lose it by failing. The railroads came the nearest to failure, and hence their rescue by the Government was the most dramatic episode. That rescue made possible an efficient supply system within the Army. But all the War Department's traffic rules could scarcely have brought about improvement if general freight had remained out of hand and uncontrolled. For this reason, the war service of the United States Railroad Administration is one of those triumphant achievements which can scarcely be overestimated.

CHAPTER XII

SPACE-SAVING IN CAR AND SHIP

NATIONAL efficiency in war is to no little extent a matter of cutting out the wastes and reforming the loose practices that spring up in normal times, when industry tends to more or less slipshod practices. The United Kingdom impressed this fact upon its people with its "doctrine of goods and services," an English household phrase during the war. Waste, of either materials or effort, however small may be the individual item, subtracts just so much from the national efficiency; and the sum of all of the wastes in a country the size of ours is far beyond what most persons would suppose. The mill hand found it hard to realize that a leaky steampipe endangered the country's safety, however slightly, when his country was under an extreme necessity to make the most efficient use of all its resources. No doubt it was difficult for a boy in war time to understand that for him to buy a pair of skates which he did not need weakened his country in its stand against the enemy. The sequence was complex; it went back to ultimate resources, ores and coal and human toil, all locked up in the toy which his dollar purchased, but which ought to have been embodied in a cannon or in shell or some other material of war. "For want of the nail, the shoe was lost; for want of the shoe, the horse was lost"—the old saw is precisely expressive of national efficiency in war. A military defeat might well be chargeable to the carelessness of the conquered people about stopping up the little wastes. This was a lesson which the American people would have had to learn thoroughly, had the war been greatly prolonged.

In rail and water transportation, there was an invisible waste which bore directly upon the efficiency of our freighting. Until the Government had called certain scientists into its war organization to analyze the subject, it is doubtful if many in the United States realized how profligate the nation had become in the use of its transportation facilities. Now and again our practices came up for comparison with the methods of the more prudent peoples of Europe—for example, in foreign trade, where our goods met those of Europe in competition—and then we heard a great deal that was not complimentary to ourselves, about, for instance, the packing of goods. Yet, even then, we did not realize that we were just as careless and unscientific about the packing of goods in domestic commerce. It is not too sweeping to assert that one of the great incidental war benefits to the United States will be the lessons learned while circumstances were forcing us for the first time to apply science to the packing and stowing of goods in transportation.

Astonishing revelations were made by the experts who grappled with waste in shipping. Probably normal industry in the United States pays $1,500,000 every day in goods lost or damaged because of improper packing and in car space wasted by uneconomic use. In war, this annual half billion, which is what it totals, represented a tremendous impairment of the nation's power. To correct the evil, the War Administration early built up an organization which, though unostentatious, worked with far-reaching efficacy. The standardization branch of the Division of Purchase, Storage, and Traffic so demonstrated its usefulness to military traffic that it has now become a permanent part of the War Department.

The benefits arising from the elimination of waste in transportation were direct and immediately effective. Employed in the old-fashioned way, the equipment of 2,500,000 freight cars in the United States was none too large for the war traffic; and indeed it was not until the Government itself operated this equipment as a whole, under management skilled in the extreme, that American rolling stock became sufficient to the needs of the country. But when—as actually occurred at one interior depot after we had begun to learn something about the efficient use of shipping space—our experts repacked into fifteen freight cars some war department property which had filled fifty, the nation was using economy which, applied to all inland commerce, would forthwith treble or quadruple our railroad equipment, without affecting the actual work of operating the trains. One can scarcely exaggerate the importance of such economy in rail transportation.

When a like efficiency in packing and stowage was applied to ocean freighting, the benefits were even more startling. Men who had grown gray in export shipping were surprised by the inefficiency which they detected when they had begun to study packing as a science. In the supply of the A. E. F. across the sea, the scientific hoarding of space by the War Department resulted, by a conservative estimate, in the saving of at least 1,000,000 cargo-tons. It requires roughly 1,500,000 deadweight tons of shipping to float such a burden. The Germans succeeded, in their two most triumphant months of unrestricted submarine warfare, in sinking just about this quantity of world deadweight tonnage. In effect, then, the brains and science of those on our docks and piers and in our terminal bases lifted that enormous tonnage from the ooze of the ocean floor and presented it to the United States, loaded with A. E. F. supplies, for one final trip across the ocean.

While the War Department was reducing the subject of packing to an exact science, the Railroad Administration was applying to all inland commerce a greater efficiency in loading cars than the country had ever known before. Intensive car-loading became the care of the car-service section of the United States Railroad Administration, whose work in this quarter decisively helped to ease traffic conditions in the spring and summer of 1918. The section had no leverage upon the commercial shipper, except persuasion; but the directions and recommendations for loading which issued from its office in Washington were generously received by industry. Records kept during the first nine months of 1918 show that the average freight car which moved in the United States in that period bore a cargo 4,200 pounds heavier than its lading in 1917.

The reform was accomplished not merely by the physical effort of loading cars more heavily, but also by changing the sizes of packages, the manner of constructing them, and the methods of packing articles within them. Tobacco, for instance, had been shipped in hogsheads too tall to be packed in double tiers inside a box car. A car loaded with tobacco in the old way contained a large empty space between the top of the layer of hogsheads and the roof of the car. On the suggestion of the Railroad Administration, the shippers adopted a shorter hogshead, and then it became possible to pack cars to the roof. The manufacturers had been shipping such things as scythes and grain cradles with the awkward crooked handles fastened at right angles to the blades. These and other agricultural implements they took apart and did up in compact bundles. The factories removed the handles from baby carriages and packed them in with the carriage bodies. They knocked down and compactly crated wagons, wheelbarrows, and other vehicles customarily shipped on their wheels. It had been railroad practice to ship such barreled goods as molasses, oils, and tanning extracts in single tiers within cars, because a heavy upper tier might crush the barrels underneath. Care in packing offset this danger by means of protecting layers of dunnage between the upper tier of barrels and the lower. These are but a few of scores of instances of the reform in car-loading.

Major General Goethals, builder of the Panama Canal, was largely responsible for the success of army packing. Having been associated with the Emergency Fleet Corporation, he understood at first hand the crisis in ocean tonnage and the necessity of utilizing to its utmost capacity every cubic foot of cargo space afloat. When he became Acting Quartermaster General, he learned something about the army packing methods then in vogue—methods inherited from the easy-going past, when shipping space seemed to be almost as limitless as the air. One of General Goethals's first acts when, as director of Purchase, Storage, and Traffic, he became responsible for all military transportation on land and sea, was to call distinguished engineers and other experts into the Army and form them into a group which became eventually the engineering and standardization branch of P. S. & T. When these experts began their study, they soon found that commercial packing practice was wasteful of space, uneconomical in the use of packing materials, such as lumber and burlap, and disregardful of the hazards incident to transportation—particularly transportation by sea, where the handling of freight is much rougher than in ordinary rail transportation. Supplies were arriving in France for the A. E. F. with packages broken, contents lost or damaged by exposure to sea water or the weather, and addresses or other markings obliterated. One of the early accomplishments of the experts was a system of standard markings, so waterproofed and otherwise applied that directions for the delivery of supplies overseas could not be lost, and so plain that they could not be misread. This change almost immediately resulted in a notable improvement in the handling of supplies at the ocean gateways to France.

It is eminently worth the reader's while, if only on the score of interest, to examine certain of the specific economies of space achieved by Major General Goethals's experts. Begin, prosaically, with hay. The A. E. F.'s requirement of hay amounted to almost a million tons a year. The commercial bale of hay, the only sort of bale that our Army had known before 1917, was of such density that a ton of hay occupied about 220 cubic feet. When our first horses and mules were shipped to France, the Quartermaster Service sent after them hay in bales of the commercial type. Packing experts, examining this problem among the first matters taken up, decided that it was possible to compress hay much more tightly than was done in commercial baling. Special baling compresses were built and installed at army supply depots; and soon we were squeezing our overseas hay down to about eighty cubic feet a ton. So compressed, the hay not only arrived in better condition—for the new bales were so tight that they were practically impervious to water—but the space required on railroad cars and in the holds of vessels for a given amount of hay was cut down nearly two-thirds. One hay ship packed in the new way could do the work of three packed in the old. With so tremendous an economy for a beginning, the packing service was at once enabled to extend its function in many other directions.

The shipment of machinery offered greater difficulty to the space-savers; yet remarkable economies followed their investigation. Each type of machinery presented a special problem, any solution of which the experts had to test by several criteria—ease of assembling when the shipment reached France; the possibility of breakage; the size of crates or boxes in relation to ease in handling. Motor trucks and camions, for instance, had at first been shipped abroad on their wheels, lashed to the decks or in the holds of cargo vessels. It was impossible so to pack these vehicles as to prevent their movement in a rolling ship; and many of them suffered considerable damage in transportation, not to speak of the loss of cargo space. At the beginning it had really been necessary to ship only assembled trucks, because of the lack of assembling facilities on the other side of the Atlantic. But soon the necessary assembling plants had been set up in France; and then the Army established shops for the proper crating of motor vehicles here, under directions laid down by the packing experts. Chief among the crating establishments was the Motor Transport Corps' Camp Holabird, near Baltimore, Maryland.

The scientific crating of standard trucks and automobiles proved to be one of the most effective space-savers in the overseas supply system. A truck which, on its wheels, occupied 1,500 cubic feet could be so condensed that the crates and boxes fitted into a space of 300 cubic feet—one-fifth of the space occupied by the assembled truck. Special crating companies and regiments, trained at Camp Holabird, became so expert that trucks moved into the camp in a steady stream and flowed out of it to the cargo piers in a corresponding stream of packed crates and boxes. Camp Holabird, when working at capacity, could crate two miles of trucks in a single day. Moreover, the crated trucks, protected from the blows and strains of movement within the holds of vessels, and also from the rust of accidental wettings or ocean moisture—for the crates were moisture-proof—reached France in better condition than those shipped on their wheels.

The crating of guns of the mobile artillery, each weapon a special problem in itself, also brought about large savings in space.

The experts applied themselves to the more compact baling of cotton. An immense tonnage of this commodity had to be shipped to the European Allies, and every foot of vessel space saved could be utilized for the shipment of other essential war supplies. By means of slight modifications of the cotton compressors in use, it was found possible, without injuring the fiber, to reduce the cotton bales one-third in bulk.

Again and again the experts studied special shipments and, by ingenuity in packing, presented fresh tonnage space to the Government. They designed an improved crating system for a consignment of 2,000 horse-drawn ambulances, thereby saving a total of 300,000 cubic feet, or about the cargo capacity of a 5,000-ton ship. They improved the crating of a consignment of 6,500 water carts, saving 279,000 cubic feet of space—a free gift to Uncle Sam of another ship for a voyage across the ocean. Airplanes required a protracted study, but the study cut in two the space previously allotted; in other words, after the recommendations of the standardization branch had been adopted, one ship could carry twice as many airplanes as it could early in the war. There had been great waste of space in the shipment of camp equipment. Camp stoves of the old type would not pack handily; it was necessary, in crating, to knock them down and nest together their three conical sections. A new stove was designed. It dismounted into three pieces which, packed flat, occupied less than one-sixth as much shipping space as the older stove. There were thousands of these stoves and literally hundreds of thousands of other articles of camp equipment; and the general average of space-saving in shipment of them, after the new packing regulations were in force, was nearly forty per cent.

Another way in which the War Department saved space in

Photo by Signal Corps

TRUCK READY FOR CRATING AT CAMP HOLABIRD

Photo by Forest Products Laboratory

THREE-WAY-END CRATE CONSTRUCTION

Photo by Forest Products Laboratory

TESTING DRUM AT FOREST PRODUCTS LABORATORY

Photo by Forest Products Laboratory

IMPROVED BOX (FOR TRENCH-MORTAR SHELL) UNHARMED BY TUMBLING TEST; ORIGINAL-DESIGN BOX BROKEN

freight cars and ships was by changing methods of manufacture. Nearly everyone has heard of the dehydration and evaporation of fruits and vegetables which was carried on during the war. This practice enabled the supply service to ship additional millions of pounds of food to the soldiers abroad in the vessel space available. Soap was dried for lighter weight and smaller volume. The reclamation of worn uniforms, both at home and abroad, was important in that it saved the space that would have been required in shipping new garments if the old ones had not been saved. Beef was boned for more compact shipment; and the adoption of the so-called shankless beef, with the hind quarter disjointed and folded in against the carcass, amounted to the addition of thousands of cubic tons of refrigeration space to our rail and water shipping equipment. Soluble coffee cut down the bulk of coffee shipments by eliminating all but the potable elements of the bean; and the practice of roasting the green coffee beans at camps and stations, both in this country and abroad, made it possible to send coffee in bulk, so that it could be stowed compactly, instead of in cans, which had to be boxed.

The huge amount of space required by the containers of tinned food was, in fact, a constant concern to the space-saving experts. Notable reforms were under way when the armistice called a halt. Already, however, we had abandoned canned pork and beans—three-quarters of the shipping weight of which is water, tin, and wooden boxes—in favor of dry beans in bulk for baking in France. This plan increased the bean supply three times with the same transportation space. Had the war continued another year, it would have seen in France a great army cannery, which would have received from the United States the raw materials of sheet tin and perishable food products in bulk, to the enormous saving of space aboard cars and ships.

In addition to these and dozens of other economies, the efficiency engineering of the War Department brought about great reforms in the standardization of supplies—changes which had the incidental effect of saving ship space. At the outset of the war the Army was using seventy-seven different kinds of tool chests. The standardization branch reduced these to seven, which were not only sufficient for all field needs, but also much more compact. Some of the new chests were but half the size of those they supplanted. On a ship that could hold 17,000 chests of the old type, it was possible to load 82,000 of the new. In one month of 1918 the A. E. F. requisitioned 30,000 tool chests. The saving of ocean freight charges, had this requisition been filled with chests of the new type, would have exceeded $1,000,000.

Electrical equipment, too, was standardized. The A. E. F. required prodigious quantities of electrical supplies. Before standardization became a fact, supply officers in France had to cable lengthy descriptions and specifications of each article required. Standardization, by giving every article a catalogue number, took an immense burden from the Atlantic cables and also eliminated errors in cable transmission. When all the electrical equipment needed by the Army had been standardized, it was discovered that the officers had virtually standardized the entire electrical equipment of the United States—an achievement which would have been deemed an impossibility in time of peace. Standardization extended also to all other army supplies. If we were to undertake another overseas war, our expeditionary forces would cable for their supplies almost entirely by catalogue numbers.

The Forest Products Laboratory, Madison, Wisconsin, an agency of the United States Forest Service at the University of Wisconsin, became an invaluable ally of the War Department in its freight-packing activities. This institution was established by federal funds nine or ten years ago for the scientific study of wood in its relation to American industry. The laboratory was of primary value to the Government in its study of wood problems connected with the manufacture of airplanes; but it rendered another important service by its expert advice on wooden packages and wooden containers of all sorts. Crating and packing had been a subject of research at the Forest Products Laboratory from the date of its establishment. When the war came, the institution was prepared to take the lead in this new industrial science. In addition to using its test rooms and its experts in the solution of individual problems, the Forest Products Laboratory set up on the university campus a training school to which the various production bureaus of the Army sent officers and men to be taught the best methods in packing. Some 400 soldiers attended this school during the war period. Much of the saving in military shipping space was due to investigations conducted at Madison.

The experts of the Forest Products Laboratory studied the packing of an army article for overseas shipment with three considerations in mind. Foremost was the consideration of strength; for, above all, supplies had to reach the A. E. F. in good condition. The second consideration was space. Every effort was made to design a crate or box of minimum size; and in this connection the experts worked out systems of packing and compressing supplies within the boxes. On one occasion an instructor at the laboratory brought before a class of ordnance officers a new box packed with entrenching tools. He allowed the students to examine the scheme of packing as closely as they wished. Then he dumped the contents upon the floor and told the students to pack them again into the box. After an hour of concerted effort had failed to get the tools back in, the instructor showed the class the system upon which his results depended; and thereafter the students themselves were able to duplicate his performance in a few minutes. The third consideration in army packages was cost. Scarcely a packing problem brought to the Forest Products Laboratory during the war missed being solved there with a stronger package, a smaller package, and a package less costly in money and materials.

The laboratory at Madison is fitted out with special testing machinery for subjecting packages to the sort of treatment which they undergo in transportation—particularly in ocean transportation, where boats must often unload in rough water, the cranes dropping their cargo nets heavily into tossing lighters. One of these machines, a great hexagonal wheel fourteen feet in diameter, with a rim eight feet wide, is called a testing drum. The interior of the rim is divided into eight compartments, each seven feet wide, with sides several inches high. The package to be tested for strength is placed in one of these compartments. The wheel, which revolves on a horizontal axis at the rate of one revolution a minute and is capable of sustaining the fall of a package weighing 1,000 pounds, carries the package upward. Near the top the package slides over the edge of the compartment and falls ten or twelve feet to the bottom, only to be borne aloft again for another fall. This process is continued until the package breaks, or "fails." The laboratory has also a smaller drum, seven feet in diameter, suitable for testing boxes up to 200 pounds. The practical value of the falling test lies in the fact that the container, loaded as it would be in commerce, is subjected to exactly such hard knocks and drops as it might receive during shipment. Another machine at the laboratory tests boxes by compression upon the corners or edges; and a fourth drops wooden containers cornerwise upon an iron floor to determine their resistance to such shocks.

The investigators at Madison have discovered that for many years American packing practice has been going down hill. The boxes and crates once in common use were much stouter than those of to-day. With competition keen in the box industry, the addition of even a single nail in a manufacturing process may put a box factory out of business. Competition has resulted in cheapening the quality of packages. Now, even one or two side-edge nails will make an astonishing difference in the strength of a box. Take for illustration a canned-food box, 21 inches by 10 inches by 12 inches, a standard size. For the best results, this box should have ten nails along the 21-inch edge; seven nails in the edge is good-quality practice. Three or four nails—the common number—are little better than none. If the ends of this box are of seven-eighths-inch material and the sides of half-inch material (the common practice), and you load the box with filled cans and drop it six inches, that single fall will likely knock out the sides if only three or four side-edge nails are used. It should stand a drop of four feet without injury. Put in five nails, and the box will stand twenty drops in the small revolving drum; put in seven, and it will stand a hundred falls, or five times as many. The reason for this increased efficiency is that five nails are not enough to transmit the elasticity of the wood in one side of the box to that in the other, whereas two more nails make a rigid connection that throws the entire spring into the wood, which then absorbs the shock. Every box car running at freight train speed has a weaving motion to and fro. If boxes are insufficiently nailed, the elasticity of the wood is not transmitted through the nail joints, and, as the packers put it, the nails simply "walk out" of the boxes. The addition of even one nail will sometimes prevent this mishap and bring a box through to its destination in good order.

As the result of the decreased quality of wooden boxes, boxes of fiber and corrugated pasteboard have been able to dispute the packing field, although these are inferior to good wooden boxes. But, having once opened the door to the competition of these materials, the wooden-box makers are now confronted with the impossibility of improving the quality of their product—a curious business situation. To improve quality means to increase expenses at the box factory, and hence to increase prices. Fiber boxes are now the product of a considerable industry, which is strong enough to usurp the field if the wooden-box makers greatly increase their prices. But the fiber or corrugated pasteboard box never would have arisen as a serious competitor to the wooden box, if the quality of wooden boxes had always been maintained.

When, at the outbreak of war, the Ordnance Bureau faced the necessity of shipping heavy supplies to France, its officers proceeded to draw specifications for crates, boxes, and other packing cases. These men knew the characteristics of metal, but not those of wood; and as a rule they specified heavier lumber than was necessary. They prescribed dimensions not standard in the box industry, and in numerous other respects departed from common practice. Sometimes they specified extra strength in places where the stress would not be heaviest. Not only were the boxes which they requisitioned less efficient, as a rule, than standard boxes, but it was almost impossible to obtain any bids for the manufacture of boxes under their specifications. When the box industry failed to bid for the contracts, the Ordnance Bureau appealed to the box experts of the Forest Products Laboratory. These men went to Washington and revised the specifications to conform to ordinary commercial practice; and thereafter the Army had no trouble in securing bids.

It seems to be the opinion in to-day's industry that strength can not be secured in a wooden box without the use of cleated ends. For years the Forest Products Laboratory has advocated the so-called "three-way end," which is one of the oldest types of box-end construction. In late years three-way-end construction has been largely discarded in favor of dove-tailed ends or cleated ends. An illustration facing page 152 shows the principle of the three-way end. In a three-way corner, each member is nailed in two directions, so that no one board can be removed unless the nails are first drawn from one side; otherwise, the board will split or break. Each piece holds the points of two nails in the grain side of the wood (nails so driven hold much better than those driven into the end grain); and four nails are driven into or through each piece, but all separated from each other.

The Forest Products Laboratory used this type of corner construction in practically all the crates and most of the boxes designed for the War Department. It is as strong as cleating, and much more economical of space. The advantage of the three-way-end box over the cleated box in space economy may be seen by taking, for example, a cubical box of half-inch material seven inches long, interior measurement. Such a box when cleated is nine inches long. With three-way ends, it is eight inches long. The cleated box occupies, then, twelve and one-half per cent more space than the three-way-end box. In boxes of average sizes, between six and eight per cent of the shipping space is saved by three-way ends. The average box maker seems to regard cleating as easier in construction, although it is not really so; and in recent years the cleated box has almost supplanted the three-way-end box in domestic commerce.

The Forest Products Laboratory brought about, over and above its general reforms, many specific benefits in our military shipping. It redesigned for the Ordnance Department the six-inch trench-mortar shell box, with savings of thirty-two per cent in lumber and nearly twenty per cent in cargo space. This incident is typical of the work of the Laboratory. In the shell box of the original design, one of the sides formed the cover. To open the box, it was necessary to take out twelve screws. In the laboratory design, the end became the cover; and it was held in place by four vertical metal straps bent down and nailed. All that had to be done to remove the cover was to cut these four straps with a clipper—an operation quickly performed, even in the excitement of battle. One box of each sort, each with a "dud" shell inside it, was subjected to 264 tumbles in the test drum. The laboratory box sustained this test without injury, whereas the shell kicked out one end of the ordnance box.

In all these stability tests, boxes were packed, whenever possible, with the contents actually to be shipped in them. Shells and grenades were shipped to Madison for these tests. In boxes designed to carry high explosives, sand or some other similar substitute, in weight equal to the powder, was used. Even articles so delicate as electric lamp globes were used in the tumbling tests, for, to be efficient, the package must protect from injury whatever contents it had to carry. The Laboratory shattered thousands of dollars' worth of electric lamp globes in developing improved types of containers for export shipment.

The Laboratory saved nine per cent in cargo space by redesigning the original box for entrenching shovels. A redesigned box saved seven per cent of the space occupied by entrenching picks. A redesigned box achieved the saving of thirty-two per cent of the space given to entrenching hand-axes.

The Ordnance Bureau designed for rifle grenades an apparently strong box which failed to give good service. The Forest Products Laboratory discovered in its tests that this was because the grenades were not packed tightly in the interior cells, which were made of partitions of corrugated board. The experts reduced the size of the cells so that the grenades would fit snugly, incidentally reducing the size of the box; and the redesigned box, although otherwise of identical construction, sustained the tumbling tests.

The Ordnance Department designed cases for Browning rifles and their accessories, the shipping equipment consisting of one box which held two of the automatic rifles and another box for accessories. The Forest Products Laboratory designed a single box to hold both the rifles and their entire equipment, saving twenty-eight per cent in material besides the labor wasted in handling two boxes instead of one. Artillery lead harness boxes were redesigned to take up less than three-fourths of the space which they had occupied. The Ordnance Department had packed ten Modified 1917 Enfield rifles in a box. The Forest Products Laboratory designed a box which held twelve rifles instead of ten and which saved thirty-five per cent in shipping room. There was waste of space in the box originally designed to hold two Browning heavy machine guns and their equipment. The Forest Products Laboratory designed a box that was just as good in every way, and seventeen per cent smaller.

Of the Army's achievements in the saving of shipping space, one of the most spectacular was that in the baling of clothing and other textile supplies at the supply base in Brooklyn. The idea of baling clothing for shipment overseas was the child of Major David Abercrombie, a name everywhere familiar to lovers of the open. As the former head of a widely known sporting-goods firm bearing his name, Major Abercrombie possessed long experience in the packing of supplies for compactness and ease of transportation; for his concern specialized

GRENADE BOXES WITH LOOSE- (ON LEFT) AND TIGHT-FITTING CELLS AFTER TEST

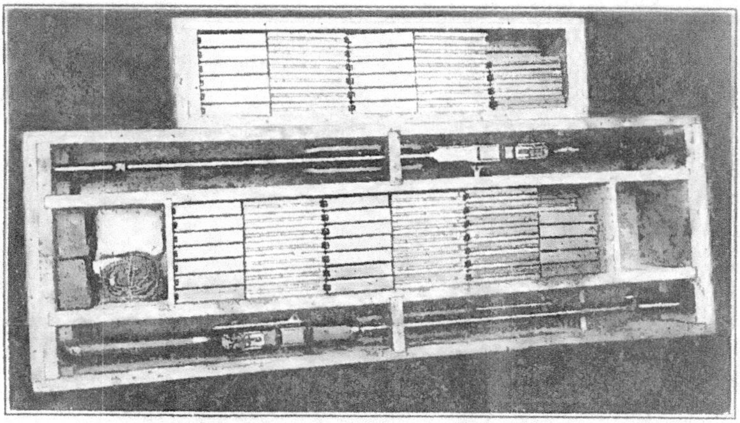

1. ORDNANCE DEPARTMENT BOXES FOR TWO BROWNING RIFLES AND EQUIPMENT

Photos by Forest Products Laboratory

2. IMPROVED SINGLE BOX FOR SAME *MATÉRIEL*

Photo by Signal Corps

AN ARMY BALING MACHINE

Photo by Signal Corps

ARMY CLOTHING BALES READY FOR SHIPMENT

in the equipment of exploring expeditions, particularly those into the heart of Africa, where all the baggage and supplies must be carried on the backs of black tribesmen. Major Abercrombie came into the army service riding a single hobby, the scheme of baling textile articles instead of putting them loosely into packing cases. He continued to ride this hobby, thinking of nothing but baling all day long and dreaming about nothing else at night, until the end of his service. The Government benefited exceedingly by his enthusiasm, for he succeeded in effecting the adoption of baling as standard army practice—a practice which, it is estimated, saved the Government during the war period close to $100,000,000 in ocean freight charges. This sum represents an enormous volume of ship space saved; how much it is impossible to estimate accurately, but it is safe to say that it was by far the largest element in the total shipping economy.

When Major Abercrombie was commissioned and assigned to service at the army base in Brooklyn in charge of the packing of supplies, the Army was shipping its textiles in large wooden cases. The quartermaster case of the specifications was of white pine lumber and weighed from forty to forty-two pounds. By midsummer of 1917 there were no white pine cases to be had—they had virtually disappeared from the trade—and the substitute case, of yellow pine, weighed from ninety to ninety-six pounds. Major Abercrombie first procured a test of baling. Its superiority over casing was so clearly demonstrated that there was no longer much question about the form in which our textile supplies should go to France. But the commercial baling machinery in ordinary use lacked sufficient power to compress the bales according to Major Abercrombie's specifications; and he had to have special machinery designed and installed at the baling plant in Brooklyn.

By the spring of 1918 the new balers were ready for work, in a wing set aside for the purpose at the Bush terminal on the Brooklyn water front. The lingering remnants of prejudice in the Army against the new system had been overcome; and thereafter baling, which had been conducted in a small way with hand compressors since October, 1917, rapidly superseded casing in the shipment of our clothing and equipage supplies, whether for overseas or from the Brooklyn base to interior points. In inland traffic the system was extended first over the East and then throughout the country, until, before the treaty of peace was signed, the Army had adopted baling in its standard specifications for packing all textiles, wherever shipped.

The list of supplies adaptable to baling is a long one. Among the better known articles of army equipment which crossed the ocean in bales, the following may be named:

Aprons	Bags, surplus kits	Mosquito bars
Blankets	Barrack bags	Stockings
Breeches	Bed sacks	Caps, winter
Coats	Bed sheets	Leggins
Drawers	Toweling	Cot covers
Overalls	Tentage	Mops
Overcoats	Mattresses	Pillows
Shirts	Mattress covers	Pillow cases and sacks
Trousers	Mosquito head nets	Caps and gloves
Undershirts	Shelter halves	

The baling of supplies was worked out as a science; a science which even possessed its own terminology. Countless experiments determined the proper way of folding each article so as to put the greatest number in a bale. The numbers packed in a bale varied widely, of course, according to the article. Only one storage tent could be put in a bale; 201 summer undershirts were compressible to the same dimensions. The ideal package for shipment is twice as long as it is wide, and a little higher than its width. The dimensions adopted for the army bale were 15 inches by 30 inches by 14 to 19 inches. With the welfare and efficiency of the stevedores in mind, army bales were usually kept under 100 pounds, but above 80. Since women were sometimes employed as freight handlers in the supply service of the A. E. F. in France, the Army preferred lighter bales to heavier ones. In a bale of these dimensions, 10 long winter overcoats could be compressed, 20 blankets, 35 olive-drab uniform coats, 45 olive-drab breeches, or 82 flannel shirts. A bale of overcoats weighed 96 pounds, a bale of blankets 83 pounds, one of uniform coats 99 pounds, one of breeches 95 pounds.

Each bale was wrapped first in waterproof paper and then in burlap. It was hard at first to get suitable wrapping paper. The paper must not only be impervious to moisture, but it must also be expansive or elastic in texture, so as to stretch without tearing at the edges where the bale is held by its steel bands. At first this paper was supplied, in a makeshift way, by crinkling paper by hand at the baling plant and then waterproofing it. But in order to obtain an adequate commercial supply, Major Abercrombie invented a special expansive paper consisting of an asphalt film between two sheets of paper, the whole corrugated. He patented this paper and assigned the patent to the Government for free use.

Experiments went on at the plant until it was scientifically determined just how much compression any given textile article could stand without injury to the fabric. Bales were usually made up of small bundles of regulation size, called "bricks" by the balers. It was found that the most garments could be placed in a bale by making up a bale of these smaller bundles. Besides "bricks," there were other constituent packages with special names. A "brick" was 14½ inches long and 7½ inches wide; four "bricks," it will be perceived, made the bottom layer of a bale. A "bundle" was 29 inches long and 14½ inches wide; it made one entire layer in the bale. A "half bundle" was as long as the bale, but only half as wide; there were two "half bundles" to a layer. A "double brick" was as wide as the bale but only half as long; two of these, also, made a layer. These forms of packages were adapted to various garments to be baled. The garments were first folded into these forms and then squeezed in a hand press until they held the bricklike shape. The bale was made up of as many "bundles," "bricks," "half bundles," and "double bricks" as it required; then it was subjected to hydraulic pressure. When such small articles as mittens or leggins were so packed, corner boards were used in the bales.

The economies wrought by baling were both numerous and large. Assuming that white pine cases had been available, the number of garments making up a 100-pound bale would have required, if boxed, 84 pounds of lumber and other casing materials. The elimination of this material effected a saving in weight of more than forty-five per cent. This weight-saving was important in the handling of supplies, particularly on trucks. It cost about $2 an hour to operate a truck. At that rate it cost eight cents an hour to truck one bale of garments. Boxed, the same number of garments, hauled at the same trucking rate, cost twenty cents; in other words, baling saved sixty per cent. Shippers reckon cargo space by ship-tons, a ship-ton being an arbitrary cubic measurement. Stevedores, in loading the hold of a vessel, could average sixty-five cased army overcoats to the ship-ton. Bale the overcoats, and they could pack 150 in each ship-ton—a space saving of sixty per cent.

Baling also saved much operating space at the shipping terminal; for, of course, a much larger storage space is required for boxing materials than for baling materials. Three cubic feet of burlap and paper will cover fifty bales of goods. To cover the same goods cased requires 620 cubic feet of lumber, nails, etc. The average cost of baling during the war was eighty cents a bale; casing the same number of garments cost $4.09. In overseas shipments, even at the low freight rates secured by the Government, each bale saved $62.11½ in ocean freight charges. Since up to March 1, 1919, the Brooklyn baling plant alone packed well over a million bales of clothing equipment for the A. E. F., the single item of ocean freight charges on clothing was reduced by upwards of $60,000,000. In addition, the saving in ship-tons brought about by this baling amounted to nearly 140,000, or the equivalent of more than 200,000 deadweight tons of ocean shipping.

Moreover, clothing shipped in this way was subject to scarcely any losses in transit. The breaking of a packing case often meant the ruin of its contents, but a bale of clothing could be dropped into the water and fished out again without damage. The compression was so tight that water penetrated the bale with great difficulty, even if it succeeded in getting through the waterproof paper covering. One of our ships containing baled supplies sank in harbor. Several days later she was raised from the bottom, and the recovered bales were found to be dry and in good condition; whereas the cases aboard the sunken ship were water-soaked and their contents ruined by muck, salt, and grease.

Baling also saved space that had been given over to the shipment of large quantities of burlap to the A. E. F. for use in making sandbags for fortifications. Each bale was wrapped in two sheets of burlap, each one of which would cut into exactly four sandbags. Nor was any of the other baling material wasted on the other side. The corner boards made dugout shingles or splints for use by the Medical Corps. The steel bands on the bales were the right size for revetment fastenings.

The proper marking of bales was a matter for scientific consideration. French stevedores might understand enough English to be able to handle our packages according to directions; but on the docks in France were Singhalese and laborers of other strange nationalities who could read neither English nor French. All our standard baled supplies were given, therefore, stock numbers, and the packages were marked by a system of Arabic numerals. These the foreign workmen on the docks could read. The system worked so well that it has now been adopted as standard. Incidentally, much study has been given to the marking of other supplies and to the correct wording for such markings. It is interesting to note in passing that the English word "fragile," derived from the Latin, is understood by folk of many nationalities, who have similar words from the same root in their own tongues.

Baling is not by any means a modern discovery. From the

most ancient times, men have shipped goods in bales; and even to-day the most primitive peoples of the earth make extensive employment of this form of packing. The Chinese are expert in the art of baling. Magnificent rugs come out of the countries north of India in bales. Boxing, in fact, is a product of European civilization. The British Army also adopted a system of baling; but the British bales were not uniform in size, as ours were. Our bales also maintained a consistent density for articles of the same specific gravity, a claim which could not be made for British bales. Moreover, American army bales were compressed to a greater density. On the whole, a comparison seems to justify the assertion that our system was scientifically superior.

PART TWO
THE PORT

Photo by Harris & Ewing

BRIGADIER GENERAL FRANK T. HINES

Chief of Embarkation Service, January, 1918–December, 1918; Chief of Transportation Service, December, 1918–August, 1920

CHAPTER XIII

A HALT BY THE WAY

ALONG the picturesque spine of the Hudson Palisades lay Camp Merritt, most beautiful of all American military posts created during the World War. Of the splendid force that upheld the prestige of American arms in France, a full quarter tarried by the way in Camp Merritt, brief guests awaiting embarkation on transatlantic ships. To more than half a million young crusaders, solemnly but eagerly setting their faces toward the east, Camp Merritt bade farewell as they departed on the great adventure, to many of them the greatest of all adventures, and the last. To an even larger number of returning heroes Camp Merritt cried "Hail!" as, with faces turned westward, they saw—nay, now all but touched—beyond the wooded horizon, the blessed vision of discharge and home and peace. In the two years between the autumn of 1917 and that of 1919, more than a million American soldiers experienced, for a few days at least, the hospitality of the camp. Camp Merritt was one of the great caravansaries on the road to France.

In the prosy chart of army organization, Camp Merritt hangs on its proper hook, a sub-branch of the New York Port of Embarkation. Tell it not to the overseas soldier, though, that Merritt was anything subordinate. To him it is a bold landmark graven in the intaglio of his war memory. It was the jumping-off place, the starting point of the crowded, excited period of embarkation.

The impressions of those hours can never be forgotten. The soldier reached Camp Merritt grimed and weary from a rail journey across, perhaps, half the continent. Merritt offered him rest just at the time when, in the stir and bustle of preparation, rest seemed impossible. To the inland youth, if he were of the lucky ones who received twenty-four hours' leave, the camp offered, too, more probably than not, the first opportunity of his life to see the great city just across the Hudson. And after that the departure—the silent march out of the camp in the dead of night, down the macadamized highway to the river, the last quarter-mile of it upon a thoroughfare that was half road and half trail, descending stiffly to the water's edge; there the landing, a platform open to the sky, lighted by incandescent clusters, its bitts mooring a small squadron of river ferry-boats, lumpish craft soon packed to their gates by the brown-clad travelers. Then the dark river, now and then streaked with undulating golden ribbons from waterside lights that grew thicker after an hour or so, when the boat left Spuyten Duyvil Creek behind and began hooting for a passage through the never-ceasing traffic of North River. Then the jutting, brick-and-stone headland of lower Manhattan, thrusting its vague silhouette against the graying eastern sky. Next the pier, tucked in between the projecting sterns of monster transports; the climb up from the ferry-boat; the echoing vault of the roofed pier, dusky in spite of its myriad white arcs; long tables spread with sandwiches and doughnuts, and Red Cross women at rolling coffee urns; inspections, inspections, and more inspections—medical inspections, equipment inspections, alienage inspections, inspections by intelligence officers—and instructions of many sorts.

Outside, the city now roars into wakefulness and a new day. The embarkation force comes on duty. Then the gangplank, the loud roll-call, the sharp scrutiny of individuals; at the top of the climb, the mail bag for the "safe-arrival" postcards and the last letters home. After that, the crowded standee berths erected in holds, companion ways, nooks, corners—any place where there had been a trifle of spare room. A peremptory order confined one to these stuffy, congested quarters during the trip down the bay, lest enemy eyes on shore or on some harbor craft mark this as a troopship. And so the chance to sleep, welcome to men drawn and haggard with twenty-four hours on their feet; the awakening to nauseous motion and the vibrant shudder of driving screws; permission to go on deck; the whip of salt wind, refreshing as a cold dip; far to the northward the faint cloud of the Long Island shore, elsewhere tumbling waters; and, from the horizon rim ahead to that of the sunset, a double file of transports, grotesquely streaked with deceptive bands of paint, escorted by flanking destroyers; dirigibles and airplanes overhead and, far in the van, a towed balloon.

Such is the typical individual soldier's retrospect of embarkation—an experience brief in point of time, yet so novel to most who underwent it that it impressed its consecutive details deeply in the memory. It began at the embarkation camp, which therefore figured as the bisector between home and overseas service.

Merritt was by no means the only embarkation camp. Also at New York was Camp Mills, originally the canvas camp of the Forty-second Division, a National Guard organization, but during the most momentous months of 1918 exclusively an embarkation camp with continuous accommodation for 40,000 transient troops. Part of Camp Upton, Long Island, was also set aside for the reception of soldiers awaiting embarkation. At the Newport News, Virginia, Port of Embarkation were two more embarkation camps, Camp Stewart and Camp Hill, as well as special camps for embarking aviators, artillerymen, and labor troops.

Of the embarkation camps, Merritt was the first established. It shared with Camp Mills the distinction of being the largest. In beauty it surpassed all other camps of any kind within the United States, not even excepting Camp Lewis at American Lake, Washington. Because of certain special functions, Camp Merritt was likewise the most important of the embarkation camps. The Embarkation Service regarded it as its model camp. To avoid the tedium and possible confusion of attempting to describe all the embarkation camps, this narrative will picture overseas embarkation as it occurred from Camp Merritt; and the reader will understand that Merritt, in most of its activities, was eminently typical.

First, something about the site of Camp Merritt—a square mile or more of characteristic New York suburban territory, not quite come to full development. The southerly pickets of the camp were two or three miles from the northern limits of the town of Tenafly, New Jersey. Tenafly is somewhat inland. Its adjacent waterside community is Fort Lee, New Jersey, so named from a fortification erected by revolutionary troops in 1776. From Fort Lee, ferry-boats run every few minutes directly to West 125th Street, New York's principal uptown Rialto. The ferry station on the New York side is convenient to the Manhattan rapid transit system.

In this New Jersey district north of Tenafly, and three or four miles back from the Hudson, the Government leased a solid rectangular block of ground, regardless of the state of development of the included parcels of land. Through a part of the site ran a splendid highway, paved with the meticulous smoothness demanded of their surburban roads by the motorists of New York. Some of the cross streets found in the camp site were paved and improved; others were merely dirt roads ending eventually, at the range of hills to the west, in rotting corduroy construction laid down originally by the troops of General George Washington. The site was a level plateau sloping slightly to the southward and broken with forests and clearings. Here and there were magnificent shade trees, oaks and elms singly and in clusters; here and there, too, were the countryseats and estates of well-to-do urbanites. At the northern end of the camp, farther away from the city, were the barns and outbuildings of occasional small truck farms.

The builders of the camp preserved its scenic beauty as completely as they could. The new dwellings there and the weather-beaten and lichened houses were alike fitted into the general building scheme as structures useful to the post—homes for nurses or officers, or headquarters for various military or welfare organizations. The axes spared the ancient trees, and the barracks were placed in the old clearings. Even some of the orchards were left, for the sake of their shade and their fruit. The planners took full advantage of the roads already

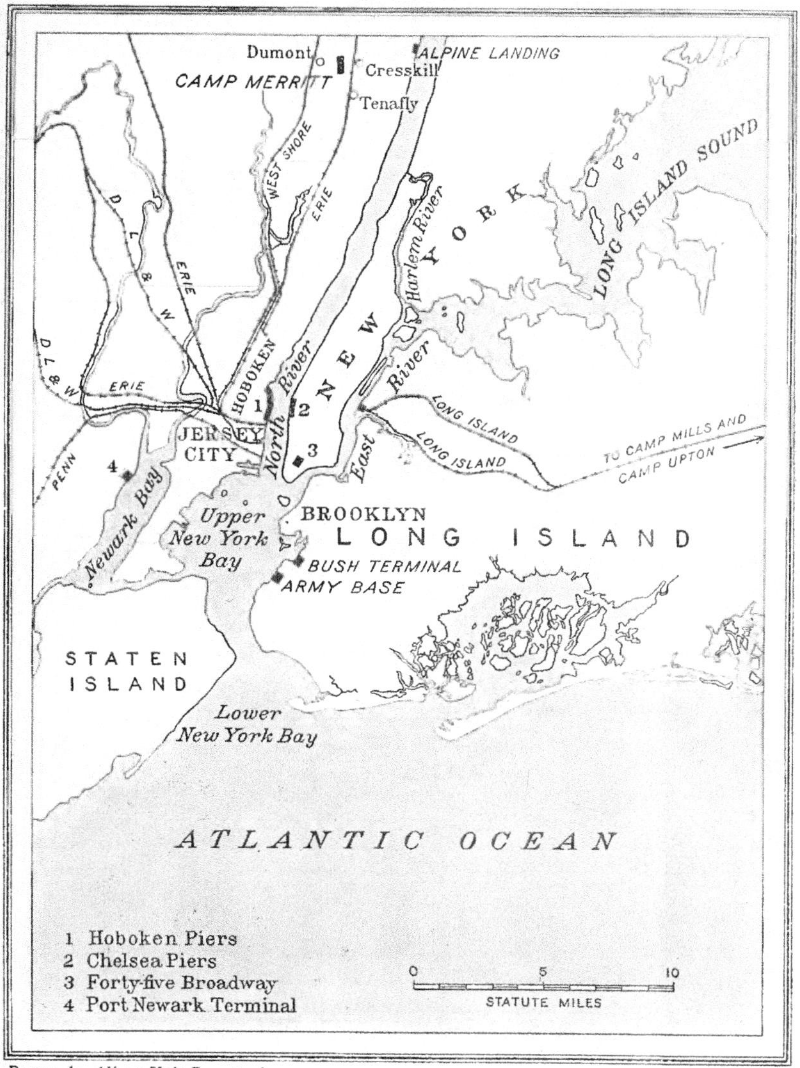

Drawn by Albert Hoit Bumstead

PORT OF EMBARKATION, NEW YORK

there, and in general built the camp to conform with the terrain as they found it. Finally, marvel of marvels, some super-officer did what no other camp quartermaster was able to do: he obtained funds wherewith to paint the camp. A battery of painting machines sprayed the pigment on the buildings, and window and door casings and eaves were trimmed by hand. When the builders finally left Camp Merritt, it stood a green and white city for 40,000 men, heavily shaded, cut by fine streets, and including among its more than a thousand structures fifteen or twenty buildings of some architectural pretension. Camp Merritt must have appeared as a refreshing surprise to tens of thousands of soldiers lately come from the usual drab, unpainted, grassless, and shadeless training camps.

Not only were surroundings beautiful, but they were of surpassing historic interest. Within the camp limits there was a spot where a handful of revolutionary patriots had been slaughtered by the mercenaries hired out by the Prince of Hesse to the Redcoat Army. The road cut in the face of the Palisades, down which a great part of our overseas army moved in 1918 to reach the ferry landing, was built originally by Lord Cornwallis in his military operations out of New York against the Revolutionists. Across this country General Washington marched and countermarched before retiring to Valley Forge. It was a region for traitors and spies during the Revolution, a rendezvous for the bearers of illicit communications. Just north of the limits of the camp Major André paid the extreme penalty for being caught within the American lines.

After the Revolution, the region outdid its native son Rip Van Winkle by sleeping for six times twenty years. Then the movies were invented, and the Palisades country became once more the scene of stirring incident and dramatic *dénouement*. Here, in close proximity to New York, were quiet rural meadows and sun-dappled woodland paths, separated by only a river's breadth from the great city. The region soon came to be a favorite "location" for the moving-picture companies whose studios were in the metropolis. Here at Camp Merritt, thousands of genuine heroes in olive drab, on the verge of departure for a sterner stage, saw mimic heroes exhibiting their prowess before the eye of the cinema. Every sunny day the moving-picture troupes were out on the roads between Camp Merritt and the Hudson River ferries; and the inquisitive doughboy caught glimpses of this strange world and duly catalogued them among his war reminiscences.

By early July, 1917, it had become evident to the War Department that the Army Transport Service, the military organization at New York which had been charged with water transportation, and which had superintended the embarkation of the First Division, was inadequate to the work ahead. The Army Transport Service had been a branch of the New York quartermaster depot. The time had now come when military transportation could no longer be regarded as an incidental quartermaster undertaking. Because of the conditions under which we fought, transportation had inevitably become one of the major military operations. America's position was unique. Other belligerents could subordinate their transportation: the fighting area was right at their doors. We, three thousand miles away, had to put transportation on a parity with the training and equipment of troops and the actual maneuvering of them in battle. The American Transportation Service, put to the test, waxed correspondingly expert and powerful; and eventually it performed the miracle which was America's chief contribution to the enduring military history of the World War—the miracle of setting an army of 2,000,000 men across the Atlantic Ocean in little more than a twelvemonth.

Upon the Army Transport Service as a foundation, the War Department erected at New York the Port of Embarkation, a distinct branch of the Army. Its commander, a major general, set up headquarters at Hoboken, at the seized piers of the German transatlantic steamship companies. The Port, regarded for the present purpose as an organization of men, was charged with the administration of all details of the embarkation of troops and of the loading of supplies which passed through New York and the subsidiary ports. These details included the direction of the army transports, owned or chartered. The Port at once made requisition for a place set apart—a camp in which overseas troops might remain while awaiting their ships. Ocean traffic was hazardous and uncertain. It was impossible to discount all accidents and delays in inland transportation. The organization could not coördinate train movements with vessel movements so infallibly that incoming troops would always reach the port at the exact moment when the ships were ready for them. There had to be provided a reservoir of men at the port—an embarkation camp.

How far short of what was actually to occur were the early estimates of the volume of traffic which would pass through New York, the fleeting weeks were soon to show. As plotted on paper early in August, 1917, Camp Merritt was to have capacity for only 20,000 men. That capacity seemed adequate. Moreover, the Port asked for no other embarkation camp—only for the one. But greater and greater grew the power of the American Army to train troops and transport them, more and more insistent the demands of the Allies for troops. Within four months the army constructors started on a project which eventually doubled the size of Camp Merritt. Camp Mills, with a capacity equal to that of the enlarged Camp Merritt, was added to the port scheme; Camp Upton, the Long Island cantonment, ceded a block of space to the Port; and still, with accommodations for 80,000 to 100,000, the facilities for men awaiting embarkation were far short of the requirements. The goal toward which the armistice found the New York Port of Embarkation striving was rest-camp facilities for 200,000 men—an equipment that would have been none too large in 1919, judged by the rate at which the troop-transport fleet was growing and by the expansion of the Government's plans for the induction and training of soldiers.

The earliest conception of an embarkation camp postulated merely a temporary shelter for the care of troops during a few hours' or days' delay in sailing. But military evolution soon gave the embarkation camp a more significant place in the scheme of operation. It became the reservoir upon which the transports drew for their passengers. When offensive operations came to an end, the plan in operation was to maintain constantly in the embarkation camps at New York a reservoir of 60,000 men. Even this number was insufficient to insure absolutely against the day when a transport might lie empty at a New York dock with no troops at the port to go aboard her. The ideal, as our embarkation officers learned by experience, was a reserve of men at the port sufficient to meet half the sailing needs of an entire month. At the height of travel in the summer of 1918, we were embarking at New York upwards of 225,000 men a month. Consequently, embarkation-camp space at New York, to have been adequate, should have given accommodations to at least 110,000 men; whereas the actual space, filled to capacity, could accommodate only about 80,000. The embarkations at that time, it will be seen, emptied the camps nearly three times every thirty days. Only the extraordinarily well-managed adjustment of inland travel to the sailing schedules prevented breaks in the flow of troops. Although the program of military expansion in 1919—the so-called 80-division program—did not call for the shipment of more than 300,000 men in any one month, there might well have arisen an emergency in which it would have been necessary to embark 500,000. To meet such an emergency easily, the Port of Embarkation at New York should have had camp space for at least 200.000 men; and this, as we have noted, was the goal toward which development was actually tending.

It was through these four or five pools, the embarkation camps, that the entire expeditionary army flowed; and hence it was here that the military authorities could most conveniently and thoroughly subject the overseas forces to a final scrutiny. A small, highly expert inspection service, working here, was able to make sure that every man who went to France went properly equipped, not only in material supplies, but likewise in physique and morale. The War Department continually issued orders to the whole Army relating to the selection

From An Official Motion Picture

OVERSEAS TRANSIENTS OCCUPYING BARRACKS, CAMP MERRITT

Photo by Signal Corps

READING ROOM IN MERRITT HALL

Photo by Signal Corps

TROOPS ARRIVING AT CAMP MERRITT STATION

Photo by Signal Corps

MARCHING INTO CAMP MERRITT

of men for foreign service and to their proper equipment for it; and it was at the port that the authorities certified to their own satisfaction that these regulations had been enforced. Here at the embarkation camps the inspection of clothing and all other equipment was most severe, and here were issued the enormous quantities of supplies that made the American Army, so far as the individual soldier was concerned, the best clothed and the most completely equipped that set foot on the soil of France. And the inspection went beyond these materials of war. It went into the soldier's health and his mental attitude. The embarkation camps became the Army's filters for straining out the dangerous alien, the spy, and the man of questionable loyalty to the United States.

The contractor started the construction of Camp Merritt on August 20, 1917. By September 17 the camp was partially ready for occupancy, and the camp commander moved in and set up headquarters. Two weeks later, various troops for the operation of the camp were in their barracks. The permanent guard regiment was the 49th Infantry, 2,000 men. In early October the camp accommodated its first transient visitors, two engineer service battalions, each of about 100 men. In November nearly 5,000 troops passed through the camp to the transports. In December the first phase of construction was completed, and the first big troop movement began when numerous organizations of the Forty-first Division moved into Camp Merritt—13,000 troops in all, of whom nearly 5,000 embarked that month.

On December 20 construction began on a large addition to the camp, designed to raise its capacity to 35,000 troops. The building operations continued throughout the winter and spring; week by week new barracks were added. In January about 14,000 troops went through Camp Merritt; in February the number was nearly 30,000; in March it was close to 45,000. In April about 47,000 officers and enlisted men experienced the hospitality of the camp. In May the transients numbered close to 50,000. In June came a great jump, almost to 82,000—a figure equivalent to the emptying and refilling of the camp two and a half times in four weeks. A soldier might remain at the camp as long as two weeks before proceeding to Hoboken for embarkation; but a commoner experience was to be there only twenty-four hours.

In June a second addition to the camp was authorized, to bring the capacity up to 45,000. This construction was completed in time to serve the demobilization.

Camp Merritt had every convenience and luxury that could be given to a military post of this sort. Its equipment included 1,302 buildings, of which 611 were two-story barracks, each with accommodations for sixty-six men. It possessed complete water, sewage, electric, and heating systems, mess halls, stores, fire stations, garages, a great chain of warehouses, a refrigeration plant sufficient for the victualing of 45,000 men, a bakery, a theatre, and also the best equipment of social and welfare buildings erected at any camp in the United States.

The camp's chief deficiency was its lack of railroad facilities. The site lay between the West Shore Railroad and the Erie, about a mile from the nearest station on either line. It was expected by the designers of the camp that one or the other of these railroads would build a passenger terminal within the camp. Because of the expense, neither road cared to undertake this, although the West Shore ran a freight switch to the camp warehouses. Troops arriving at camp by train had to march, sometimes, two miles to reach their barracks.

There was, however, another means of transportation between the camp and the piers—the natural deep-water artery of the Hudson. On the river near the camp was Alpine Landing. Before 1918 only occasional excursion steamers touched there, or lighters carrying supplies to the New Jersey communities. The landing was simply an open, unroofed platform, but it could be used as a ferry slip. In May, 1918, the Port of Embarkation chartered several ferry-boats and began to operate them between Alpine Landing and the piers in North River, a distance of some twelve miles. This innovation put upon the troops the burden of marching about three miles from the camp to the river landing; but it saved the Government $75,000 or more, which was something, and, what was more, it saved the terminal railroads at New York the work of transporting hundreds of thousands of troops for this short distance. The metropolitan railroad managers testified that, but for the Camp Merritt ferry service, they would have had to devote hundreds of passenger cars to the transportation of troops on this route and perhaps, at times of greatest military activity, to curtail by so much the normal suburban traffic facilities at New York.

Soldier-welfare work took a prominent place in the activities of all the embarkation camps. Here at the last stopping place on American soil it was particularly desirable that the troops should experience in concrete form the approval and gratitude of the American people. Here was the final opportunity to inspirit them and to build up their morale. Not that their courage was flagging; but they were facing invisible dangers amid strange surroundings, at sea—and most of these lads had never seen the ocean. Every effort was made to put the men on the ships in a buoyant mood that should carry them triumphantly through the perils of the deep.

There was another phase of welfare work at the embarkation camps—a pathetic phase. It was here that thousands of mothers and fathers, wives, sisters, brothers, and sweethearts said their farewells to the boys who were going across. It became a special problem to take care of these relatives and friends, who were often simple country folk, unacquainted with travel. At Camp Merritt this work was the concern of the Young Women's Christian Association, of the War Camp Community Service, and of several other welfare organizations.

One of the most beautiful buildings at Camp Merritt was the Y. W. C. A. Hostess House just inside the main entrance. The house was built in the Elizabethan style, with broad gables and wide verandas. The furnishings were comfortable, even luxurious. The Y. W. C. A. opened this building for service late in September, 1918. Prior to its dedication the organization had occupied two other buildings at the camp.

The first of these was a fine old countryseat at the top of a hill near one of the entrances; it commanded a beautiful view of the valley east of the camp and of the Palisades beyond. This mansion was opened as a Hostess House on the first day of December, 1917, and within a few days it began receiving an unexpected number of visitors. The villages of Dumont and Cresskill, New Jersey, near Camp Merritt, possessed no adequate hotel accommodations; Tenafly was several miles away and accessible only by automobile and by infrequent trains; the journey into New York from the camp, by a combination of automobile, ferry-boat, and rapid-transit train on the island of Manhattan, was calculated to dismay a country woman unused to travel. Often the visitors arrived at camp late at night. The Hostess House was never so full that a bed could not be provided for one more; and the Y. W. C. A. workers secured hundreds of additional rooms for visitors in the surrounding communities.

Those who made the pilgrimage to the camp were for the most part women—women of all ages, from immature girls to wrinkled and white-haired grandmothers—and often they were alone and in need of protection. Farewells were usually courageous, but almost always they were deeply emotional. Some came too late to see their soldiers. Some found the trip more costly than they had expected, and were stranded at the camp. The Hostess House relieved such distress—with sympathy for the sorrowing and with money for those who needed it. One bitter day in winter a father and mother who had traveled all the way from Montana reached the Hostess House, only to learn they were two hours late: their son's unit had departed. In dumb disappointment they shouldered their way out again into the storm. Another, a wife who came to say good-bye to her husband, found him dying of pneumonia in the camp hospital.

The work of finding men in the shifting population of the camp, that they might come out to meet their visitors, was a difficult task. It required at the Hostess House a proficient information bureau in touch with the ever-changing transient personnel. During the period of embarkation the Hostess House at Camp Merritt located over 75,000 soldiers and summoned them to meet their friends; and it served at least 200,000 meals to those who claimed its hospitality.

Another out-of-the-ordinary department of welfare work at Camp Merritt was that conducted by Merritt Hall, the enlisted men's club. This fine structure, with its furnishings, was the gift of Mrs. Wesley Merritt in memory of her late husband, General Merritt, after whom the camp was named. The building's low ceilings and comfortable furnishings gave it a snug, homelike atmosphere. The camp chaplain operated the club. In it the American Library Association established a camp library of over 20,000 volumes, the shelves of which lined a large, sumptuous room equipped with easy chairs, broad tables, numerous reading lamps, and a great fireplace that blazed hearteningly in cool or cold weather. In the club one could buy refreshments, stationery, stamps, and other small accessories. Within the building were a large cafeteria, a billiard room with eighteen tables, reception rooms and parlors, and—a most popular installation—a modern soda-fountain. Did the men appreciate the club and take full advantage of it? The hardwood threshold at the front door was worn completely through and replaced three times.

These two institutions, the Hostess House and the enlisted men's club, Merritt Hall, were typical in scale of much other welfare work at Camp Merritt. The American Red Cross found the embarkation camps convenient for the issue to our troops of the garments made for their comfort by the women of America. At Camp Merritt the Red Cross built several structures, including a headquarters building and a convalescent house. The volume of supplies given out by the Red Cross is indicated by such items as approximately 105,000 sweaters, 110,000 knitted helmets, 80,000 mufflers, and 150,000 pairs of socks.

In Camp Merritt was also the Officers' Club, a gift through the Red Cross of Mr. Cleveland H. Dodge. This structure, built at a cost of $43,000, was an important adjunct to the camp equipment, for it furnished hotel accommodations at cost to thousands of overseas officers awaiting embarkation at Camp Merritt. The Young Men's Christian Association maintained five buildings in the camp and a large force of secretaries to assist in their operation. The War Camp Community Service was also active; and so were many other organizations, including the Knights of Columbus, the Jewish Welfare Board, and the à Kempis Society, an organization of Catholic women which maintained a visitors' house at the camp entrance.

Camp Merritt has played to the end its part in the World War. The last divisional troops have passed through the camp in demobilization. Yet, come what may, the camp is not to disappear utterly. The leased site will not all be turned back to the owners. Merritt is to remain a historic spot. The citizens of Bergen County, New Jersey, have purchased, at the main street-crossing within the camp, a circle of ground. This is to be dedicated to posterity as a public park; and in the center of it will rise a monument, to which in future years the half million or more veterans of the World War who remember Camp Merritt fondly may return, to revive again in memory the days when they reached the port and set their faces toward France.

CHAPTER XIV

IN CAMP MERRITT

IN days before the World War came to upset the accustomed plan of life, one of the popular American institutions was the personally conducted tour to the Holy Land or to some other foreign clime attractive to the sightseer. The personally conducted tour was born as an institution about the time when the Steamship *Great Eastern* was the marvel of the world, and it continued to flourish and to put red-letter summers into the lives of school teachers, college professors, country bankers, and inland doctors and philosophers until the Year of Grace 1914, when certain events in Europe combined to place foreign travel among the hazardous occupations. Since then and until almost the present day, personally conducted touring has suffered a complete hiatus, although, as the hordes of Americans flock to see the battle fields of France, there are indications that the custom is about to be renewed, and on a scale which will make all such former excursions appear as mere trickles of humanity.

To join such an entourage, one had only to answer the advertisement, send on his lump-sum appropriation, and report at the dock at the appointed hour; and thereafter he was entirely in the hands of the touring agency, which supplied quarters to him aboard the specially chartered steamer, arranged for his meals and hotel accommodations on the other side, looked after his baggage and found it if it went astray, procured guides, supplied overland conveyance wherever it was needed, protected the traveler, shepherded him hour by scheduled hour over the rigid route which he had paid to follow, and finally delivered him at the promised minute back upon his native shore. During all this time the tourist was completely divorced from the ordinary vexations of travel; he needed only to keep eyes and ears open and drink in experience.

Now the Transportation Service aimed to be to the Army what Mr. Cook's admirable organization is to the amateur globe-trotter. How well it succeeded, these pages are designed to show. But the Service went even further. Certain of its activities at Camp Merritt and the other embarkation camps were much beyond the province of the mere travel bureau. Let us see.

Suppose that when the expectant tourist reached the steamer pier and prepared to go aboard, the gentlemanly conductor took him aside and addressed to him words such as these: "Oh, my dear sir! You are not at all properly dressed for our expedition. That suit you have on may be comfortable in this climate, but it is not the thing for where we are going. Just step with me into our equipment room here and let me make a selection for you. Now, here is the latest creation in tropical wear—the ultimate in coolness, yet of a sturdiness to withstand the wrinkling and hard usage of travel. And that hat! It will never do. Here is a cap for steamer wear; here a cork helmet, with neckshade, for use on the burning Syrian desert. I see you have no sun-umbrella. Here is a serviceable one. We stop for six hours at the island of Corfu; there are said to be brigands there. This pistol is issued to you for your protection." And so on, *ad lib.* Surely the traveler would think that this was the ultimate refinement in personal conducting.

Well, it was precisely what the Transportation Service did for the Army. The Service attended completely to the Army's travel, and in addition it saw to it that the Army was properly dressed and equipped for the service which it was to perform at its overseas destination. Such an evolution came about naturally enough. The embarkation camps were virtually turnstiles through which the entire overseas force had to pass man by man. Where, if not here, was the War Department to verify the uniformity and adequacy of its Army's equipment? The duty of making this verification fell to the Transportation

Service, the organization which was on the ground. A corollary of inspection of this sort is the issue of new supplies in place of those condemned; therefore the Embarkation Service, over and above its transportation duties, came to be a great supply agency, and its embarkation camps became supply centers. In addition to ordnance and quartermaster goods of the familiar sort, issued to take the place of those rejected in the port inspections, the embarkation camps furnished to the departing hosts those new and novel articles which belonged exclusively within the overseas equipment—the spiral puttees, the steel helmets, the gas masks, the trench knives, the entrenching kits, Sam Browne belts for officers, and other articles, all of which our overseas soldiers saw for the first time after they reached the ports. The issue of supplies to soldiers became, then, one of the principal activities at each embarkation camp.

The Army did not rely upon its field commanders for that uniformity in equipment which the A. E. F. exhibited when it debarked in France. Ostensibly, the war department regulations required the correct equipping of units before they left their training camps for the ports,—in fact, the sailing orders which went out from the Embarkation Service specified just what equipment each soldier should have,—but these regulations were intended only to remove some of the burden from the supply depots at the port camps. The forces scattered throughout the United States proved to be unable to equip themselves according to regulation. Some of the commanding officers at the interior camps were careless. Others were so engrossed in purely tactical duties that they could not spare the attention which the equipment required. Still others, compelled to leave camp on short notice, could not obtain the necessary supplies in time. Moreover, the regulations were continually changing. The growing experience of the A. E. F. command was constantly reflected in its cabled additions to and emendations of the specifications for overseas equipment. The busy field officer in this country could not reasonably be expected to keep up with this incidental development. The scientific arrangement was to let the field commanders do the best they could, and to create at the ports a compact inspection and supply service which could put into instant effect any change in the equipping regulations.

The embarkation-camp supply service became exceedingly skillful in its work. One of its inspectors could glance at a soldier and tell wherein his clothing and equipment were deficient or excessive. A few seconds' appraisal of the contents of knapsack and kit was sufficient. Moreover, the inspections were conducted right in the issuing warehouses, with the necessary supplies at the inspector's elbow. Red tape was cut. Substitute supplies were issued on the spot forthwith, and the lines of soldiers filed through the buildings almost without halting.

Often an outsider might have been puzzled by the quality of the garments discarded in this final inspection. The rejected clothing sometimes seemed as good as new. But the inspector was constantly influenced by a consideration that entered into so many supply questions in 1918: the inveterate necessity of saving ship space. If a soldier took to France a uniform which would not give him practically 100-per-cent wear, he would call for a new uniform sooner, and the replacement uniform would have to cross the ocean in a cargo hold, taking up its bit of the sorely needed room. Better to send the new uniform on the soldier's back in the first place—such was the law of the port. Clothing discarded at the embarkation camps went through salvage, eventually to be reissued, but it never crossed the ocean. We sent only new clothing to the A. E. F.

The quartermaster department of Camp Merritt grew from a small beginning. At first its personnel numbered only a few officers and enlisted men. In the summer of 1918, when the embarkation movement was at its height, the Camp Merritt quartermaster supply enterprise gave occupation to nearly 2,500 officers and men, who constituted more than half the post's permanent garrison. These troops occupied an entire block of the camp. Camp Merritt was then easily the foremost quartermaster post in the United States. A score or more of quartermaster warehouses testified to the magnitude of the undertaking.

It was in vain that the quartermaster organization at Camp Merritt besought the port authorities for more time in which to inspect and reclothe the overseas soldiers. The officers asked that incoming organizations be allowed to remain in camp at least three days before being called to the piers. Such waits were often out of the question. The fleets of army transports and allied transports at our disposal had grown to such proportions that convoys were sailing almost daily, and hungry piers were devouring troops as fast as railroads could haul them in. The average stay at the embarkation camp became, during July and August, 1918, twenty-four hours or less. Troops were dispatched from Camp Merritt at the rate of 4,000 a day. The camp's constant population of more than 35,000 transient troops changed completely at least three times every month. The arrival of any large contingent of troops at Camp Merritt meant for the supply companies a stretch of work that often ran thirty-six hours without a break. During this activity there was no sleep for anybody; the supply troops scarcely paused to snatch hasty meals.

The inspection of clothing was rigorous. If a uniform showed that it had already given more than a third of its normal wear, it was rejected, and the supply service passed out a new uniform to the soldier and ordered him to turn in his old one. So with shoes; so with shirts and underwear. The inspection extended to the soldier's reserve supply of garments in his barrack bag. Every article of apparel had to measure up to the requirement—less than 30 per cent of a theoretical 100 per cent of wear. The inspectors used numerous tests of their own to determine the degree of a garment's deterioration. The camp inspection condemned all campaign hats and canvas leggins of domestic issue, regardless of their condition, and substituted the overseas caps and woolen puttees.

To thousands of overseas men in the busy months of July and August, 1918, Camp Merritt was a rest camp in name only. Many of these men never saw the interior of the camp barracks. The convoys were in port awaiting their loads, and the port reservoirs of overseas troops had dwindled to nothing.

At such times the reception committee at the railroad station escorted the troops directly to the warehouse section of the camp. There they received their new supplies and donned them as they passed through. Then out of camp they marched, and down the road to the ferry landing at Alpine.

Since the embarking troops usually left Camp Merritt between one o'clock and six o'clock in the morning, the work of issuing supplies to soldiers frequently continued until long after midnight. If the incoming troops filled the issuing warehouses, the work of supply overflowed upon the camp streets near by. The quartermaster troops loaded great motor trucks with clothing and stationed them at intervals along the street. Lined up at the curbs were files of men, their baggage unpacked and spread out upon the pavement before them. Down the lines passed the inspectors. If a pair of breeches appeared to be worn, the inspector grabbed them by the seat and gave a mighty jerk that told whether the seams were strong. Other tests were equally rough-and-ready. Supply squads accompanying the trucks brought out new garments as they were needed, and the soldiers were ordered to strip where they were, regardless of the temperature or the hour, and put on the new apparel. Such street scenes were common enough in Camp Merritt during the summer of 1918.

Both the quartermaster and the ordnance depots at the camp were often hard put to it to procure supplies for the hundreds of thousands of troops whom they were called upon to equip. Often the long warehouses were bare of numerous necessities, such was the heavy drain upon them of the unprecedented rate of embarkation. But somehow, by hook or crook, the camp always managed to get what was needed. If the freight deliveries fell down, the supply officers called upon the express; if that were too slow, they brought clothing and other equipment by motor truck directly from factories or from the supply bases at New York. They became desperate men. On one or two occasions they located shipments of supplies *en route* to other military posts, and reached out and diverted these consignments into Camp Merritt. Month after month they fought in

Photo by Signal Corps

TRUCKS UNLOADING QUARTERMASTER SUPPLIES IN STREET, CAMP MERRITT

Photo by Signal Corps

TROOPS DRAWING SUPPLIES IN STREET, CAMP MERRITT

Photo by Signal Corps

QUARTERMASTER WAREHOUSE, CAMP MERRITT

Photo by Signal Corps

MOUNTAINS OF EMPTY BOXES SHOWING TREMENDOUS ISSUE OF SUPPLIES TO OVERSEAS UNITS AT CAMP MERRITT

this fashion for supplies, and they always got them. Not a man departed from Camp Merritt deficient in his equipment, although, of the 570,000 who passed through the camp, at least half required new clothing, an even greater fraction new ordnance equipment, such as knapsacks and belts, and every single man such novel accessories as overseas caps, steel helmets, trench knives, and gas masks.

Late one afternoon when 8,000 infantry replacement troops were in Camp Merritt, equipped and ready to start for the piers that night, orders came that thereafter no infantry were to embark without kits of entrenching tools. The port supply service was at its best in such an emergency. At the time of the order there were no entrenching kits in camp. Less than twelve hours later the 8,000 soldiers boarded the ferry-boats, every man with a kit. The tools had been freighted out from the city on motor trucks. In one respect only was the equipment below inspection par: the entrenching spades and their handles were not assembled and packed in the canvas covers. But each soldier had orders to attend to this job during the two-hour ride down the river.

The ordnance depot at Camp Merritt supplied not only personal articles to the troops, but also a considerable amount of the ordnance equipment of field organizations. It was at the camp that many divisions met for the first time the Browning machine gun. The Camp Merritt ordnance depot also shipped supplies directly overseas to the A. E. F. Among these export shipments were thousands of sets of artillery harness. The depot eventually came to be a concentration point for light ordnance supplies of many sorts, which it distributed to training camps and posts in the eastern part of the country. The depot also sold to officers pistols and other ordnance accessories.

When the overseas travel was at its peak, the embarkation camps were insufficient to accommodate the whole movement, and in the emergency the War Department utilized some of the eastern training camps as temporary embarkation camps, sending troops from them directly to the piers. Camp Meade was so used when the Seventy-ninth Division moved from it to the Hoboken water front in two days, an episode already narrated. Camp Dix, in New Jersey, became at times an emergency embarkation camp; and posts as far outlying as Camp Devens in Massachusetts and Camp Lee in Virginia served on occasions in the same capacity.

Whenever outlying camps were used in embarkation, the Port sent its inspection and supply officers directly to them to take charge of equipping the troops. Once or twice there was a direct shipment of troops from Camp Lee to the ships at New York. In each of these instances inspectors and supply officers from Camp Merritt went to Camp Lee in advance of the movement. They saw to it that each soldier, when he stepped aboard the train at the camp, wore new or practically new clothing and carried his "tin hat," gas mask, and other paraphernalia which the overseas man commonly never saw until he came within smelling distance of salt water. The same procedure was followed at Camp Meade, at Camp Dix, or at any other temporary embarkation camp.

One immediate outgrowth of the tremendous supply business at Camp Merritt was its correlated activity, salvage. The garments and articles of equipment abandoned by the overseas troops still had much wear left in them. No interior camp reclaimed clothing on anything like the scale of salvage at Camp Merritt. At the camp itself there was a large reclamation plant, including a shoe repair shop and power mending-rooms. But in the summer of 1918 this plant was buried deep under discarded equipment, and it became necessary for the camp to send reclaimable shoes and apparel to many contract factories in New York, as well as to army salvage plants throughout the East. Camp Merritt operated a large laundry at Hoboken. Before the armistice this laundry had washed and cleaned more than a million articles of clothing.

The reclamation of ordnance material was also an extensive project. The ordnance depot at Camp Merritt maintained a repair shop in which armorers restored broken rifles and pistols, other men mended haversacks, ration bags, and similar equipment, other squads repaired and cleaned mess pans and cups, knives, forks, and spoons by the hundred thousand, and still others reclaimed entrenching tools and bayonet scabbards.

The ordnance activities at Camp Merritt required five large warehouses for the accommodation of supplies. The quartermaster depot operated eighteen warehouses, three of them for the issue of clothing, shoes, and equipage, one for the exchange of clothing and shoes, two for receiving and shipping Q. M. supplies of all sorts, and the rest for storage. Two entire storage warehouses were filled with shoes, one with underclothing, three with uniforms and overcoats, one with gloves, one with leggins and raincoats, and another with blankets. Between October 11, 1918, and November 11, the busiest month of all, the camp supply service issued over 1,800,000 articles of clothing for overseas men.

The salvage of food at Camp Merritt was extensive. Organizations moving into the embarkation camp often brought their own food supplies with them. Troops traveling on the railroads were required in most instances to feed themselves *en route;* and it was usual for an organization, on starting out from its training camp, to supply itself with subsistence not only for the rail journey to the port, but also for its stay in the embarkation camp. Unaware how long they should have to remain at the port, some of these organizations greatly overstocked their messes. But no food except the emergency ration was carried aboard the transports (the Navy or the foreign steamship companies fed our soldiers at sea) and therefore the overseas units left the excess food supplies in their kitchens.

When an organization moved out of Camp Merritt the camp authorities promptly sent men to the vacated barracks to put them in order for their next guests. In the house-cleaning which followed, the camp troops discovered great quantities of stores, principally of mess stores in the kitchens. Tens of thousands of pounds of meats, fresh or preserved, and hundreds of thousands of pounds of bread, potatoes, sugar, soap, and many other staples were taken from the vacated kitchens and turned in for salvage. Nearly all of this was saved. Such of it as could not be used currently at the camp was placed in storage for reissue as required.

Photo by Signal Corps

TROOP BAGGAGE ARRIVING AT CAMP MERRITT

Photo by Signal Corps

SALVAGING CLOTHING DISCARDED BY EMBARKING TROOPS, NEW YORK

Photo by Signal Corps

OVERSEAS TROOPS ARRIVING AT CAMP MERRITT

From An Official Motion Picture

EMERGENCY RATION TO BE CARRIED ON BOARD SHIP

CHAPTER XV

PERSONALLY CONDUCTED

PUTTING troops through the embarkation camp and aboard their ships in good order and with due attention to the government records was an intricate business. The celebrated one-armed paper hanger and his equally assiduous *confrère*, the cranberry merchant, were gentlemen of leisure compared with the commanding officer of a large body of overseas troops from the time he stepped on his train at the cantonment until he settled down with a sigh in his transport stateroom, ready at last to let somebody else do the worrying. Suppose we glance at a few of the manifold duties which destroyed his reputation for geniality.

In the first place, there was the matter of baggage—not the personal baggage of the soldiers, for they carried that in their hands and on their backs, but the property of the organization. There were its animals, horses and mules; its field kitchens and mess equipment; its tentage; its trucks, camions, carts, automobiles, motor cycles, and bicycles; its headquarters equipment of filing cases, typewriters, paper records, and the like; the trunks of its officers; its machine guns, often its artillery; its field telephone sets and other signaling equipment—to name only some of the things that had to be taken along. In the early months, organizations carried the apparatus for their own entertainment, cleaving even to such items as pianos and billiard tables. And you could be sure that, in outrageous contravention of the orders in such cases made and provided, the men of the companies would manage to slip into the consignment sundry dogs, goats, and other mascots in whose mediatory powers they trusted. It may be whispered by the way that the transportation officers often developed a strategic blindness when such contraband cargo was being loaded.

All these impedimenta had to be packed according to one set of regulations and painted and marked according to another. They had to be billed through in correct form to designated terminals—artillery to Newport News, trucks and other heavy equipment to the freight piers in Brooklyn or Newark, trunks, clothing, bed rolls, and other checkable baggage to the passenger piers in North River, records and headquarters office equipment to the embarkation camp, and so on. If any of the baggage were lost on the way to port—and with the innate perversity of luggage, some of it was tolerably sure to go astray—it had to be located and brought in, and in a hurry, too.

A pretty problem, this matter of handling organization baggage and freight *en route*. An even prettier one for the stevedores at the port; for the shipment together of nondescript impedimenta—tons and tons of it, with any large organization of troops—made difficult, if not impossible, the balanced and economical lading of vessels. Moreover, as often as not the organization and its heavy equipment failed to come together at the debarkation port in France; some divisions in the Expeditionary Forces never did recover the freight on which they had expended so much effort prior to embarkation. For almost as long as we had had an Army, it had been entrenched in army tradition and custom that a military organization must carry its freight with it wherever it went. But under the exigencies of 1918 and the incessant hammering of the embarkation forces, that particular tradition finally crashed to the ground. There finally went forth the revolutionary order that units setting out for France should take with them only their personal belongings and their business records, and that all the field and camp equipment required in France should be furnished them upon arrival, from the bases and warehouses of the A. E. F. Services of Supply.

Thus, with one stroke of the pen, the War Department thrust aside this traditional lumber that was cluttering our overseas transportation. The innovation instantly removed from the shoulders of the commanding officer of field troops the burdensome detail of looking out for much freight and baggage *en route;* it saved him the vexation of losing his organization's possessions entirely; and, above all, it made possible the shipment of divisional *matériel* to France in bulk, to the great conservation of vessel space and the equally great simplifying of expeditionary embarkation, for now the shipment of men and that of property could be handled as separate tasks. The change came too late in the game (only a few weeks before the armistice) to cut much figure in our transportation history, but its marked advantages would have been evident in another year of fighting.

The transportation of baggage was only one of the embarking commander's cares. He was confronted with an appalling volume of work after he reached the embarkation camp. He had to make up strength lists of the units under his personal command, see that they obtained their assignments to quarters aboard transports, work out the passenger lists for each vessel occupied by his troops, compile and bring up to date the service records of every single man in the command, ascertain that the necessary inspections were carried through, keep an eye upon the supplying of his men with overseas clothing and equipment, put in force the police regulations designed to prevent the escape of information to the enemy, send on advance parties to occupy the transports, detail troops to clean up and bring along essential property overlooked in evacuation of the quarters in the embarkation camp, report to the port commander, formally visit the camp commander and the commander of the transport which was to carry him across, supervise the movement of his command from the embarkation camp to the piers, superintend proceedings on the piers, turn over to the Port the service records of deserters, men absent without leave, and men taken sick or rejected for physical disability at the last moment, do half a hundred other required things—and attend to all of them within, sometimes, a few hours. To be sure, the commanding officer was surrounded with aides and assistants who looked after the details; but the final responsibility was upon him alone. There was a current impression that the C. O. was likely to be a bit irascible during embarkation time.

The War Department did not expect the field commander to know how to clothe and equip his men for foreign service, and no more did the Embarkation Service expect him to become learned in the technique of embarkation. In point of time, embarkation was, after all, but a slight incident in the career of the overseas unit in 1918; and it occurred only once in that career. Yet the details of embarkation were myriad and devious. Left to himself (as in our previous overseas expedition), the commander at best could only have floundered through the procedure of going aboard ship; at the worst, an undirected embarkation would have resulted in a fearful confusion. Accordingly, the official travel bureau—that same transportation organization which had already made smooth the trip to the port and was now about to clothe the commanding officer's men in correct garments and hang at their belts or harness to their shoulders each item of the regulation equipment—again stepped forward and, all through this trying time, stood at the commander's elbow, his mentor, guide, philosopher, and friend.

The Port set forth on paper a detailed system covering every step of the progress of an organization from the moment it detrained at the embarkation camp until its last man had boarded ship and the transports had cast off their lines and backed out into the stream. This was printed in small type, with plenty of bold-face emphasis, upon thirty-six pages bound into a booklet entitled *Embarkation Regulations*. The Port placed a copy of this pamphlet in the hands of every overseas commander a few days before he took train for the seaboard. It expected him, through perusal of the regulations, to become familiar with, though not adept in, the procedure at port. No doubt they dismayed him, these formidable requirements. But he need not have worried: experience had built up in Camp Merritt, and in the other embarkation camps as well, expert corps ready themselves to attend to the more technical preparations. The camp administration soon discovered, in fact, that the transient organizations could not prepare their sailing records and attend to other embarkation details with uniformity; whereupon the permanent command at Merritt took these responsibilities into its own hands. It came to pass eventually that about all the visiting commander had to do was to stay by and sign his name on the dotted lines.

In some of the Latin countries, where they make more of an art of courtesy than we do, there is an official of the government, a sort of assistant to the foreign minister, who goes by the rather awkward title of Introducer of Ambassadors. It is his duty to meet the distinguished guests of the nation at the border, see to their comfort and entertainment, pilot them safely through the rocks and shoals of official etiquette, and eventually bid them Godspeed as they depart. Camp Merritt had its official "introducer," except that, instead of one of him, there were several, all attached to an important branch of the camp command. They were the camp liaison officers.

The Commander of the Port fixed the date for the arrival at New York of an overseas organization. At the same time he sent to the embarkation camps a copy of the order. If the organization were assigned to Camp Merritt, the liaison office of that camp at once took cognizance of the fact. It attached one of its liaison officers to the organization, and it was to be his exclusive care during its time in the camp. He was expected to follow the inbound travel of the organization by means of information which he was able to obtain from the Camp Merritt agent of the troop-movement office, and thus to ascertain the exact hour and minute of its arrival. Meanwhile he inspected the camp quarters assigned to the organization and had them made tidy. When the headquarters train pulled into the camp station, the liaison officer was there to greet the commander. He escorted him to the quarters which he was to occupy in camp and showed him the sections to be occupied by his organization. He accompanied the visitor to headquarters and introduced him to the camp commander. Thereafter, throughout the stay of the organization in Camp Merritt, the liaison officer called upon the troop commander at least twice every day, smoothed out his difficulties, put him in touch with the various camp activities, saw to it that the camp organization was not neglecting him in any way, and in general made the visit of the commander as pleasant as he could.

One of the extremely important duties of the commander of an overseas unit resting in Camp Merritt was to compile the service records of all the soldiers to be embarked. The Embarkation Service made it an inflexible rule that no man could go on board a ship unless his paper record was complete. For each soldier there had to be a minimum of eight cards and records; there might be as many as twelve. The eight required cards and records were as follows:

(1) The soldier's service record, showing the organizations with which he had served from the date of his entrance into the military service;

(2) His individual equipment record, showing what supplies had been issued to him;

(3) His pay card, showing what money had been paid him by the Government;

(4) A record of his pay allotments;

(5) His application for War Risk Insurance, or his waiver of it;

(6) His qualification card, showing his individual assets in technical skill and mental characteristics;

(7) His so-called "locator card," giving his name, his army serial number, his rank, and the organization in which he served; and

(8) A card showing his sick and hospital record.

Additional documents might show a court-martial record, or certify scores made on the rifle- or pistol-ranges.

At this point we seem to hear the reader snort disgustedly, "Red tape!" Red tape, yes; but red tape without which the efficient management of a great army would be impossible. In 1918 the Army was spending money by billions of dollars, and tens of thousands of officers had the handling of these funds. Without the system of checks and balances which the public knows as red tape, there could have been no assurance that fraud and graft would not run riot. Unfortunately, humanity is frail before temptation. The elaborate paper work of the Army was performed largely to guard against losses of public money through unfaithful stewards. Take, as an instance, the individual soldier's pay card. Without the institution of the pay card, the opportunities for pilfering from the public treasury would be enormous. Let the critic put himself in the place of a disbursing officer. Would he pay out money to a soldier who possessed no pay card, knowing that, if the man had no right to the money, he, the payer, would be held personally accountable for it? If a soldier asserted that the Government owed him money, his pay card was *prima facie* evidence as to the truth of his claim.

The soldier's service record was his military pedigree. The military authorities were accountable to the American people for every one of the millions of boys in the Army. The service records were the Army's means of keeping track of its individual members. In the transition from America to France there was endless opportunity for men to become lost from their organizations. The insistence of the Embarkation Service that no man should sail unless his paper record was complete was the Army's best safeguard against losing track of its members. The soldier-member of a traveling unit did not carry his record on his person; it was filed, with those of his comrades, in custody of the unit's adjutant. If for any reason the soldier were detached from the command, it then became the duty of headquarters either to hand him his record or to forward it to the organization which he was to join. The Embarkation Service required these records to be instantly available on the piers, and even aboard ship until the vessels had left the docks. It might be, and often was, necessary to take a man off a ship at the last moment. With the company files at hand and open, the ship did not need to be held while a hunt was instituted for the man's service record. A service record for every man embarked, and no service record taken overseas without a man to show for it—such was the rule at the port.

The compilation of service records, however, gave little trouble to a well-administered organization whose records were kept up to the mark at all times. To such a unit the work at the port became merely a check for accuracy. A more laborious duty was that of making up the so-called "assignment list" of troops ready for sailing. This list named all the units of the organization to go, specified the character of each (infantry company, ambulance company, supply company, or what not), and gave the number of men in each unit. The prime requisite for this list was accuracy, for passenger space on the transports was figured down to the individual soldier. Upon this list the port officers based the assignment of ship-quarters to the organization.

Each organization prepared three passenger lists: one for commissioned officers, field clerks, nurses, and civilian employees of the Army, another for non-commissioned officers, and a third for enlisted men. Officers were arranged on the list by consecutive numbers, according to rank, No. 1 being the ranking officer to embark on that particular ship. The pleasantest quarters aboard the transport fell, naturally, to those of highest rank. Non-commissioned officers were arranged on their list according to grade and likewise numbered consecutively. Enlisted men, however, were arranged in the passenger lists by squads, on the basis of the usual company formation; and thereby hangs an interesting tale.

At the outset the enlisted men were arranged alphabetically on the passenger lists. The checking officers at the gangplanks put the men aboard individually. When a name was called, the bearer of it stepped forward and repeated it distinctly. The embarkation officer then turned to the company commander, who stood just behind him, and requested the man's service records. These were taken from the files, which had been brought to the foot of the gangplank. The embarkation officer then checked the soldier against his record and passed him through. Now, in however good formation a company might approach the gangplank, when men were called out under this alphabetical system the formation was soon torn to pieces, and confusion ensued. In fact, the disorder was so unmanageable at times during the early months of embarkation that several wives of enlisted men and non-commissioned officers, having somehow gained entrance to the pier, managed to slip on board the vessels while the attention of the embarkation officers was directed elsewhere, and actually accompanied their husbands to France. Moreover, the alphabetical system produced a slight delay just after each man had stepped forward, while his officer was searching through the index for his service record.

It was for these reasons that the system was changed. The men were arranged on the passenger lists according to company formation, No. 1 on the list being the first man in the first squad. What was equally important, the company's service records were taken out of alphabetical arrangement and rearranged to correspond to the passenger list, so that each man's record was on top when he responded to his name. This reform saved time, did away with confusion, and enabled the unit to move leisurely and in unbroken formation up to the gangplank, as if feeding into a mill.

Each passenger list was headed with the name of the embarking organization, including company and regimental designations, the organization's item number (a point to be explained fully later on in this account), the name of the transport or commercial steamer, the date of sailing, and the port whence the ship was to sail. And, whether the passenger were a major general or a buck private, the list impartially showed:

(1) His family name, followed by his Christian name and middle initial, and his army serial number immediately under his name;
(2) His rank and corps;
(3) His organization;
(4) Whom to notify in case of emergency (name in full);
(5) Relationship to him of the person to be notified; and
(6) Full address of the person to be notified.

The Port required of each organization five copies of all its passenger lists, signed by the company officers who had made them out. These lists were the object of more care than was put into the compilation of any other records at the port, for it was imperative that they correspond precisely to the actual embarkations. No American troop transport was sunk on its eastward voyage; but, had one been, the War Department would have been able to make public the exact list of those on board, in the full assurance that the list would contain not one name too many or too few.

It early transpired that visiting organizations were not putting into their sailing lists and passenger lists that uniformity which the port authorities desired; and consequently Camp Merritt's own staff undertook to see to it that these documents were accurate and in proper form. From beginning to end of the heaviest overseas movement, the personnel adjutant of the camp and his force of assistants checked up the thousands of visiting troops against their service records and their sailing lists and passenger lists, and finally turned these papers over to the troop commanders to be signed and forwarded to the port authorities. Fifteen men at Camp Merritt did nothing but compare company rosters and service cards. Each soldier who embarked wore suspended from his neck two metal identification tags bearing his army serial number, so that if he lost his life, either at sea or in battle later, his body could be identified. From the first of June, 1918, until the armistice, Camp Merritt maintained a detachment of twenty-four men who did nothing but stamp these identification tags.

The Government took it for granted that the spies and agents of the German Government would make supreme efforts to gain military information at the embarkation camps. Through these camps passed every soldier of the A. E. F. An effective enemy intelligence service, if it had focused at these points and obtained complete information of arrivals and departures, could have learned the exact strength of the American forces in France. Moreover, here was the natural source of information on which the enemy might have based submarine attacks on our convoys. Therefore the home Army raised its thickest and highest walls of silence about the embarkation camps; and within them ruled the most severe censorship known in the United States.

The Government did not carry its protective measures to the extreme of depriving overseas men of brief leaves of absence while their units were in the embarkation camps; but not more than one-fifth of any transient force was allowed to be on leave at any one time. Soldiers going from Camp Merritt to New York City were cautioned not to confide in strangers, even though no one in the organization, not even the commander himself, knew the exact hour when the unit would be summoned to the piers. Incoming private communications were not censored; but the camp intelligence service maintained a strict censorship over all outgoing mail and telegrams. To make assurance doubly sure, the intelligence officers also watched the post offices and telegraph stations in the communities adjacent to Camp Merritt. In the commercial offices the telegraph operators themselves acted as volunteer censors. Officers and enlisted men at Camp Merritt were forbidden to file messages at telegraph offices outside the camp limits, and the telegraph companies instructed their agents in the vicinity of the camp not to accept messages from anyone in uniform.

An important phase of intelligence work at the embarkation camps was the so-called counter-espionage. This, as applied to the Army, was the system employed to discover possible spies within the military service. A soldier's loyalty to the United States might be under suspicion; but, unless he committed some overt act that fastened guilt upon him, the Army felt it to be safer to leave him in the service, where he would be under constant surveillance, so long as his unit remained within the United States. As was natural enough in an army of nearly four million men, thousands were reported as being enemy spies. The Intelligence Service of the Army painstakingly investigated every such report. Most of the suspects were merely innocent victims of the inflamed suspicion of their fellow soldiers. But now and then a report seemed to have some basis. When the hour for embarkation came, the Army took no chances with men to whom attached a shadow of legitimate suspicion. Even though espionage could not be definitely proved against them, they were not permitted to cross the ocean. It was here at the embarkation camps that such men were finally weeded out.

The American Army numbered among its members in 1918 tens of thousands of the citizens and subjects of other countries, an appreciable fraction of them natives of Germany and Austria. The Selective Service Law expressly exempted enemy aliens from military service, a provision which had the effect of forbidding their induction. But some of these enemy aliens, notably those from Alsace-Lorraine, were eager to fight against Germany. Others, though they had never taken citizenship, were thoroughly loyal to America. Moreover, the first inductions of National Army troops came before our formal declaration of war against Austria, with the result that many Austrians wore the American uniform. After Austria became an enemy, the Army ordered the release of unnaturalized soldiers of Austrian birth. Yet many of these soldiers were from subject countries which had been oppressed by Austria, and some of them resisted discharge. The draft boards let in a few Germans. Many of the Austrians managed to stay in the uniform. Moreover, no legal provision forbade the voluntary enlistment of enemy aliens.

The Army had to handle, also, the problem of the aliens from neutral or co-belligerent countries. With the co-belligerents we negotiated reciprocal treaties which permitted us to draft their nationals for service in our Army. Neutral aliens could, by declaring an intention to become American citizens, render themselves liable to military service. Many co-belligerent and neutral aliens were in the volunteer organizations.

These neutral, co-belligerent, and enemy aliens came to the ports in the ranks of the overseas column. There the enemy aliens, no matter how bitter their hatred of their native countries, faced an insurmountable barrier. We might permit them to serve as American soldiers within the United States, but we would not take them to France. The Government made no exception to this rule. Moreover, other aliens were in a sense volunteers; they might at any time apply for discharge and obtain it. The Government, because it wanted to remove from such men the disability of alienage, simplified the naturalization process, set up naturalization courts at the army camps, and made it possible for aliens to gain American citizenship quickly.

One of the most exhaustive inspections at the embarkation camps was that which probed the citizenship of every man who applied for overseas transportation; and here the work of the naturalization courts became especially heavy. The court at Camp Merritt was established in the spring of 1918. In its first month this court naturalized 2,461 aliens, whereas only 513 aliens to whom the subject was broached refused American citizenship. From the records of that month we select for tabulation a few interesting and significant figures:

Country	*Naturalized*	*Refused Naturalization*
Great Britain	378	10
Russia	415	69
Italy	593	140
Germany	2	1
Austria	15	4
Belgium	24	1
Denmark	44	3
Sweden	74	47
Norway	54	19
Netherlands	22	1
Mexico	6	8

In the autumn of 1917 and the early part of our first war winter, the troop convoys sailed infrequently from our coast. Two or three weeks, or even a whole month, measured the intervals between "sailing days" at Camp Merritt. But when spring approached in 1918 and ship after ship joined our transport fleet, and especially when the British threw in their great mass of passenger tonnage, the departures grew more frequent, until, by midsummer, "sailing day" at Camp Merritt recurred every forty-eight to seventy-two hours. Those were the times when visiting troops scarcely saw their barracks, when holidays and leaves of absence were unknown to the permanent camp staff. Each departure represented to the camp a victory won, a great task accomplished; yet the last column had scarcely swung out upon the road to the river landing when Camp Merritt rolled up its sleeves, spat on its hands, and, with the marching in of a new organization from the railroad station, tackled another job just as hard.

There was little sleep for the overseas soldier on the night before he sailed for France. The convenience of the Port demanded that all troops to be embarked in a single day be on the piers by eight o'clock in the morning, when the force of pier inspectors and gangplank checkers came on duty. The checking of men on board the ships was exacting work. No man could continue at it long at a stretch and remain accurate. The Port conducted embarkation, then, in the morning and early afternoon, when the checkers were fresh. An embarkation took from Camp Merritt between 8,000 and 12,000 men, and the movement of these troops from the camp to the piers consumed about five hours; so that much of the trip inevitably occurred during the bleak period between midnight and dawn.

The practice at Camp Merritt was to start the departing troops out in columns of 2,000 or 3,000 men—enough so that each column fully loaded one ferry-boat. The columns were dispatched at half-hour intervals, the first ordinarily at one o'clock, the second at half past one, and so on through the four or five columns; the last one seldom left camp later than half past four. When troops had reached the camp late, the activities in preparation for their departure often continued up to the minute of the command to fall in. It was from three to four miles from the camp to Alpine Landing, according to the location of the troops' quarters in camp; so that a column was on the road to the river for at least an hour. Assuming that the ferry-boat was loaded in half an hour, it was half past two or, more likely, three o'clock before the first boat started down the Hudson, to be followed at half-hour intervals by the

Photo by Signal Corps

AN ARMY NATURALIZATION COURT

Photo by Signal Corps

REST ON ROAD TO ALPINE LANDING

Photo by Signal Corps

TURN IN OLD CORNWALLIS ROAD DESCENDING HUDSON PALISADES

Photo by Signal Corps

TROOPS AT ALPINE BOARDING FERRYBOATS FOR TRANSPORT PIERS

others. The ferry-boat took two hours to make the run to lower Manhattan; so that the first troops reached the piers shortly before five o'clock, and the last ones arrived by half past seven.

In dispatching troops to the piers by ferry, the Port violated the military principle that units should always travel together, instead of split up and mixed with parts of other units. When the columns were formed in the camp, they were made up of platoons or other fractional detachments of two or more units. A regiment would find itself divided up among the departing columns. By this system the embarkation authorities gained efficiency at the piers. If a ferry-boat had carried an entire regiment, it would have debarked its whole load upon the single pier to which was moored that regiment's transport. Under such a plan, the pier in question would have been congested with work, whereas the other piers would have been idle until the following ferry-boats arrived. As it was, each ferry-boat carried parts of several units. The first boat down in the morning touched at the northernmost pier and discharged a few of its passengers, went on to the next and put off some others, and continued down the river in this way until it had distributed detachments to all the piers which were embarking soldiers that day. The next boat delivered more men to all the piers, and so on until the last ferry completed the quotas. Thus all the piers began work practically at once, and there was congestion nowhere.

When overseas troops marched out of Camp Merritt, that faithful cicerone, the liaison officer, who had been their conductor while they were in camp, marched with the men over the miles to Alpine Landing. First to greet them when they arrived from their training camp, he was last to bid them the camp's official good-bye. As the frothing water widened between him and the downbound galleon of the commutation routes, now translated to so strange a trade, he stood on the landing alone, a living symbol of the new order in military travel.

CHAPTER XVI

CASUALS

THIS narrative has been preoccupied thus far with such activities of Camp Merritt as were typical of what went on at all the embarkation camps. Camp Merritt, however, was unique in one important particular: it was the one camp, so far as the New York Port of Embarkation was concerned, which prepared for embarkation and sent on the road to France, rejoicing or sorrowing, as might be. that orphan of the military service, the casual.

The casual surprised the American Army mightily. It was not prepared for him, had not taken him into account. He appeared suddenly and unexpectedly, like a new baby in an already overcrowded family; and, having appeared, he proceeded to grow prodigiously in number and to make lusty demands for food, clothing, and a place in the military organization. Above all, he demanded travel, although there had been no adequate provision for his travel. Consequently, his presence in the scheme of things put a heavy burden upon the Port of Embarkation, which finally solved the problem by placing upon Camp Merritt the onus of arranging for the voyage to France of all casuals.

The Army had always had, to be sure, a few casuals—always, that is, since the Spanish War. The name, however, had never struck the public ear, or become so familiar as it now is. It was an obscure word, even in American military terminology; an esoteric word of the narrow circle of quartermaster officers who managed the transportation of troops to the Philippine Islands. There were always a few casuals going to and fro between the Philippine Islands and the United States or to the Panama Canal, traveling as individuals or in little groups. They had constituted no problem whatever to those who were then in charge of military transportation. Naturally enough, then, when the Army was devising its great system for transporting the expedition to France it gave scant consideration to the casual; indeed, it all but left him out of the equation. Judging from the past, there was no reason to suppose that the number of casuals would bear any greater proportion to the body of organized troops than it had borne in our Army prior to the World War.

The Army was caught quite unprepared for what actually occurred. As early as the autumn of 1917 there were indications that the transportation of casuals might become a problem of sorts; for it was then that the Army began sending overseas unattached officers and small groups of specialized troops for particular service with the A. E. F. The officers were usually experts of one sort or another—men taken from civil life and commissioned in the Army. The inherent nature of the A. E. F. required it to be a unit almost as self-contained as an entire nation. Whatever kinds of specialists, particularly of the industrial sort, a whole nation required for the balanced maintenance of its life, the A. E. F. also needed. The earliest departures of this sort to France involved little bands of telegraph operators, squads of men experienced in cold and dry storage, detachments of stevedores, companies of cooks and bakers for distribution among the permanent posts and camps which the A. E. F. was beginning to build, officers schooled in railroading or general construction, and the variety of experts required by the Services of Supply. These soldiers traveled, not in regular organizations, but as individuals or groups under separate orders. They were, in short, casuals.

The increasing number of them presently began to add a vexatious amount of work to the duties of the Embarkation Service. There was almost as much preliminary routine detail about the overseas passage of a single casual officer as for that of a whole company of organized soldiers. It was easier to manage the embarkation of a division of line troops than to put through the Port of Embarkation a single trainload of casuals. The Port managed somehow to get through the earliest weeks of overseas movement without having to devote a special organization to the handling of the casuals; but by early winter, when the casual travel had increased rather than diminished, the regular port organization could no longer bother with this extra toil added to its general routine. It was then that Camp Merritt undertook to provide hospitality for the casuals, put them through the embarkation mill, and then hand them over to the receiving system of the A. E. F.

Practically all of the earliest casuals were, then, officers and enlisted men ordered to France on special assignments. By December we had placed in France an expeditionary force of appreciable size; and the sequel of its embarkation was a trickling stream of casuals of a new sort—stragglers, men who, for one reason or another, had fallen behind their columns on the march to France.

For a multitude of reasons, men were left behind when their organizations embarked on the ships. They might have been actual cowards and deserters; they might have been absent without leave when their units entrained at camp or embarked at the port, though guiltless of intention to desert their organizations; they might have overstayed their leave and then feared to return, dreading punishment; they might have lost their way while traveling to meet their units at designated points; or they might have missed trains, or been left sick in the hospitals, or found themselves cut adrift and lagging behind through any one of a dozen other combinations of circumstance. Innocent or guilty, culpable or merely unlucky, one and all eventually came to Camp Merritt, where the special organization checked them in, assigned them to quarters and messes, tried and convicted or acquitted the prisoners, prepared service records for those who had none, organized them all into groups, provided leaders for them, put them on the transports, and shipped them every one to France.

No sooner had the stream of stragglers through Camp Merritt begun to grow in volume than a new factor complicated overseas transportation: the A. E. F. began calling for replacement troops, and these replacements began arriving at the port. This was in January, 1918. At that time Camp Merritt had been only a few weeks engaged in handling, salvaging, as it were, and forwarding the human detritus of the on-moving host. The replacements, too, were military mavericks, unbranded, completely nondescript except for the single item of their past training. They came simply as infantry replacements, artillery replacements, engineer replacements—human plaster and putty for plugging up the chinks and gaps in the wall of men which the United States was erecting on the frontier of civilization. In number they were legion, these replacements; but they were without formal organization, usually without officers, and, at the time of the most desperate demand for troops, without discipline, as the Army defines discipline. Sometimes they came to port in their civilian clothes, directly from the draft boards. They fitted into no transportation scheme, and the Port did the convenient thing—sent them to Camp Merritt to be treated as casuals.

Camp Merritt solved the problem of the casual by creating a casual camp—a camp within a camp. This was a spontaneous evolution from the conditions. The first casuals of the straggling type began reaching the Port of Embarkation along in November, 1917. As they reported, the authorities sent them to Camp Merritt, the only port camp in existence at that time. But little attention was paid to them. The camp, with an insufficient force, was struggling to work out a new science. These estrays of the service were outside the routine line of progress, and no one was charged with responsibility for them. As a matter of form, they registered themselves at camp headquarters, which assigned them in a random way to barracks wherever there chanced to be vacant beds, and then promptly forgot them. There was no officer to see to it that they were fed, none to concern himself with whether they were properly dressed—no one to keep track of them. They were absolutely footloose, overrunning the place as camp followers and supernumeraries. They ate wherever they could find a mess that would extend hospitality to them. Many of them owed their predicament to their roving dispositions. These continued to follow their predilections, making of the camp only a sort of general headquarters where they could sleep on occasions or obtain a free meal whenever the spirit moved them to drop in. All in all, these earliest irregulars were having a fine time—largely at Uncle Sam's expense—without a thought for the service which they might some day be expected to render. In only one respect was their plight unfortunate: lacking place in a recognized organization, they became strangers to the monthly pay check.

Eventually the subject forced itself upon the attention of the camp administration. Headquarters, consulting its books late in November, found 500 men registered as casuals. Where were these men? How were they faring? What were they doing? Nobody knew. Headquarters decided to set apart several barracks for them, appoint an officer to be their general guardian, maintain some sort of restraint upon them, and eventually form them into companies for shipment to France. Cold weather had set in. Many of the casuals were clad in thin cotton uniforms, and there was no regular system whereby they could be provided with suitable clothing; consequently, a supply officer was added to the casual organization. The haphazard method of feeding the casuals in such messes as were able to take care of additional appetites was found to be insufficient; four camp kitchens were set apart for casuals and regular messes arranged.

This makeshift arrangement proved unable to subject the men to a proper restraint. On the morning after Christmas, 1917, the camp rolls showed that 760 casuals of the straggling sort had reported at Camp Merritt. That morning the camp authorities attempted to secure a detail of seventy-five men for a special job of work at the casual barracks. The officers called 760 names, but only twenty casuals answered "Present": the rest were still celebrating the holiday in New York. This trifling episode convinced the authorities that they had a difficult problem on their hands. Thereupon they formally established the overseas casual camp of Camp Merritt. An officer with the rank of lieutenant colonel was placed in command and assigned to the task of devising a system for the care of casuals. This he did (1) by occupying a definite section of Camp Merritt and forming around himself a regular headquarters organization, with adjutant, personnel officer, supply officers, medical officers, a police force, and the like; and (2) by organizing the casuals into companies of sixty-six men each, placing these companies under competent control and discipline, and otherwise forging the material of a system which in time became an important link in the transportation chain.

Overseas casual companies were numbered consecutively, starting with No. 1. By the beginning of 1918, stragglers were daily reaching Camp Merritt, either on their own initiative or on that of military organizations throughout the country, in numbers ranging from a dozen to a hundred. On January 5, 1918, the first eight overseas casual companies went to Hoboken for embarkation. Each company numbered fifty men. From that time on, the casual camp steadily grew, until eventually it became the largest single activity within Camp Merritt. Eventually some 50,000 stragglers came to the camp, joined casual companies, and voyaged to France. The apex of this movement was September, 1918, when nearly 10,000 casuals of the drifting sort boarded the transports.

One interesting oddity of the situation was that, until after the armistice, Camp Merritt never possessed formal authority for a permanent staff at its casual camp. It took Washington a long time to realize the great volume of casual travel and the difficulty of managing it. The casual-camp commander, however, could not operate without assistants; and he succeeded in obtaining some sort of left-handed permission to go among his transient guests and impress such help as he needed. More than one casual officer, having bidden his friends a proud good-bye and received their felicitations upon his wonderful opportunity to see service in France, found himself, to his surprise and intense disgust, detained at Camp Merritt and assigned to a prosaic job in the administration of the casual camp. The same fate overtook enlisted men *en route* to France as casuals on special orders. From such material the camp command picked and chose. It seized stenographers and field clerks and inspectors, all of whom supposed they were to join the A. E. F.

At one time, late in January, 1918, a company of sixty cooks, newly graduated from an army cooks' and bakers' school at San Francisco, reached Camp Merritt in response to overseas orders. The A. E. F. had at this time a serious shortage of cooks, but the casual camp at Camp Merritt had a more serious one. There were 2,000 casuals in camp, but scarcely any cooks at all. The camp command did not send the sixty cooks to Hoboken. It assigned them to the new kitchens in the casual camp. This virtual kidnapping roused wrath across the water, and for a time it seemed that Washington might step in and order the men sent on their way. But just then a providential epidemic of measles broke out in the casual camp, and one of the cooks contracted the malady. This unlucky circumstance—unlucky for no one but the victim—permitted the camp authorities to quarantine the entire outfit without violating either orders or their own consciences. By the time the quarantine had expired, Washington had forgotten about the case; and those sixty cooks—who, if they live a thousand years, will never be able to refer to the affair in gentle terms—failed to see the green shores of France. They remained to the end, serving up slum and canned Willie to the army casuals.

The Camp Merritt casual camp was well established in January, 1918, when the first replacement troops began moving into the port. The replacements resembled casuals in that they were often incompletely organized, even without officers. Yet before they could embark they had to have their complete service records, sailing lists, and passenger lists. It was convenient for the Port to treat them as casuals, and the casual camp at Camp Merritt seemed to have been made expressly for them. On no day up to the armistice were there fewer than several dozen detached and casual officers at Camp Merritt awaiting embarkation; and oftener there were several hundred. The first plan adopted by the casual camp in its treatment of replacements was to select overseas casual officers to put in temporary command of replacement units. These emergency foster-fathers were expected to attend to the embarkation records and shepherd their flocks across the ocean, after which the replacements went into the fold of the A. E. F.

The plan had, however, no great success. Many of the casual officers were men newly taken from civil life, without military experience of any sort, little familiar with army paper work and still less familiar with the handling of troops. The regular officers stationed at Camp Merritt obtained much covert amusement by watching ex-college professors or office men trying to line up companies of replacement troops and march them to their places in the embarkation columns. Casual officers, drafted for this work, soon gave way to permanent block commanders within the casual camp, each block commander having jurisdiction over the replacements within his own section. He prepared the records for the men in his block and, later, conducted them to the pier. But he did not embark with his protégés: at shipside he turned them over to some of the casual officers on the same transport. These officers might be ignorant of military procedure, but they could master the regulations for the conduct of troops on shipboard, they could put the men through the various boat drills and fire drills, and in the event of a torpedoing they could superintend the abandonment of the ship.

Less than 4,000 replacement troops embarked in January. Thereafter the number steadily grew, until in the summer of 1918 as many as 50,000 a month went aboard the transports. As the war progressed, the tendency was toward a greater proportion of replacement troops.

At length there came about a drastic change in the method of transporting overseas the casuals of the vagrant sort. For a long time the military machine on this side organized them into mixed companies and forwarded them to France in the expectation that each man would ultimately catch up with the organization which had left him behind; but this system only made confusion in our French ports. The A. E. F. was on strange soil, and the whereabouts of units and organizations were often not accurately known to the administrations of the debarkation ports. Hundreds of casuals became lost entirely in France and never found their original companies. In fact, after a time the A. E. F. abandoned the attempt to forward stragglers to their own units, and treated them all as simple replacements. The debarkation-port officers ascertained from each casual what his training had been, and then sent him to any organization which happened to need soldiers of his type.

It was this same simplified system which, by official order of the War Department, was adopted in the United States in July, 1918. This order required the Port of Embarkation to treat as replacements all casuals on this side of the water. As these men reached the casual camp, they were sorted according to their training and placed in skeleton replacement companies—infantry casuals in one company together, medicals in another, quartermaster casuals in a third, and so on. If overseas organizations reached Camp Merritt deficient in strength, the camp drew upon these casual replacement companies to bring their numbers up to par. Such a system put many of the casuals into regularly commanded ranks, and incidentally made it easier to pass them through the process of embarkation. Still, with overseas laggards arriving at Camp Merritt at the rate, sometimes, of 1,000 a day, the organized units embarking at the port could not use all the casuals: and, whenever the skeleton companies reached full strength, the excess supply embarked simply as replacements.

When a casual soldier arrived at Camp Merritt he reported at headquarters, registered, and received an order which assigned him to a certain block in the casual camp and provided him with a bed in one of the barrack buildings. His first visit was to the supply warehouses, where his clothing and personal equipment were brought up to the overseas standard. Then he was assigned to a skeleton company in the arm of the service to which he belonged. His immediate officer in the block then prepared his service record—brought it up to date if it were deficient in that respect, or made it out anew on the soldier's unsupported say-so if, as was often the fact, he had lost it. A record of the latter sort was regarded as a temporary record, to be used only until the permanent record could be located, and was so marked.

The equipping of casuals was a tremendous job. Most of them reached camp without adequate equipment; many required completely new outfits. One entire warehouse at the camp stored nothing but ordnance and quartermaster supplies for casuals and replacement soldiers.

It taxed the ability of the casual camp command to feed its guests, largely because their number fluctuated so capriciously. Organized troops brought their own mess organizations and, when they moved into camp, simply occupied kitchens and mess halls in their areas and proceeded to subsist exactly as in their training camps. But the floating population of the Army, the casuals and replacement troops, possessed no mess sergeants or kitchens or cooks; and Camp Merritt itself had to provide for feeding them. One week there might be as many as 20,000 casuals of all sorts at Camp Merritt, and the next week only 1,000. The camp mess department had to be extremely flexible. Casuals arrived at all hours of day and night, often unheralded. The casual camp operated an all-night kitchen to feed hungry arrivals between sunset and morning. Whenever any large detachment of casuals or replacements reached Camp Merritt, cooks were sent immediately to the section assigned them, and their next meal was ready at the appointed hour. When the camp was filled to capacity it took 130 men to do the cooking for the casuals. The subsistence office of the casual camp operated sporadically more than 150 mess halls.

In the spring of 1918 a third class of men began accumulating in the Camp Merritt casual camp: soldiers rejected by the Port for overseas service because of alienage, physical disability, or other reasons, including the cryptic one, "for the good of the service." These men the camp distributed from time to time among cisatlantic posts where they might be serviceable.

Camp Merritt maintained a card-index record of every man who passed through it. The camp personnel officer had charge of this activity, which in the summer of 1918 kept three shifts of men feverishly occupied making out cards day and night. The card record brought to the camp, incidentally, a duty which it had not foreseen. Soldiers were often careless about letting their friends know their whereabouts. All during the overseas movement the War Department in Washington was flooded with inquiries about soldiers. Washington could follow the movements of organized troops overseas, and it maintained an extensive branch to answer questions from the anxious relatives of men in the regular units; but Washington did not attempt to keep track of casuals. Instead, it referred to the port headquarters at Hoboken and Camp Merritt all questions relating to them. Consequently the camp, which handled the overseas passage of all American casuals except the relatively few who embarked at the independent Port of Embarkation at Newport News, Virginia, became a central bureau of information about casuals. The camp set up a separate office to answer queries, whether official ones from the War Department or unofficial ones from private citizens.

Casual officers arriving at the Port of Embarkation of New York might accept the hospitality of Camp Merritt if they chose, as many did. Often there were, at one time, 300 casual officers in Camp Merritt awaiting embarkation. To these the camp served as a hotel. They were assigned to quarters and to messes, and they paid for their entertainment the one dollar a day collected from each officer at any army mess. The officers' club at Camp Merritt was extensively patronized by these casuals.

The concentration of all New York casuals at Camp Merritt involved the camp in a related enterprise which grew to be one of its most interesting and important activities. The camp became the single agency for collecting and forwarding overseas all deserters and men absent without leave from organizations which had proceeded to France through New York and its subsidiary ports. Under the system finally adopted, all embarking organizations surrendered their claim to any men left behind for any reason whatsoever, and passed to Camp Merritt the jurisdiction over such men. Technically, then, a deserter from the Nth Infantry became, after the Nth had gone aboard its ships, a deserter from Camp Merritt; and in the eyes of military law a man absent without leave from the Blank Engineers, after his regiment had departed, was absent without leave from Camp Merritt. It became the duty of the camp to apprehend these delinquents and restore them to the military service.

Deserters and A. W. O. L.'s became so numerous that in April, 1918, Camp Merritt established within the casual camp a special stockade for the detention of these men. The prison camp thereafter expanded until it embraced sixteen barracks, each accommodating sixty-six men, and four kitchens and mess halls, the entire block surrounded by a barbed-wire fence charged with electricity. Just as the overseas casual camp was a camp within a camp, with a complete camp administration in miniature, so the stockade was a still smaller camp, self-contained, within the casual camp. The stockade, at its largest, was none too large to accommodate the prisoners who were brought in or who surrendered themselves, although the procedure was so speeded up that the average prisoner remained in the stockade only six days, in which period he went before the court-martial and was tried, convicted and sentenced, or acquitted. The stockade, accommodating as it did nearly 1,000 men, was emptied and refilled once every week, to reduce its transactions to equivalent terms. The stockade administration came to include twelve officers and about 250 enlisted men. Not all these troops were required at the stockade to manage the prisoners: some of them spent much of their time traveling about the United States, bringing in deserters and other prisoners accountable to the A. E. F.

It was possible at Camp Merritt to get some idea of the extent to which desertion made inroads upon the ranks of

America's World-War Army. Camp Merritt did not, to be sure, come in contact with desertion from units stationed within the United States; but it is reasonable to assume that desertion was greatest among troops facing the ordeal of actually meeting the enemy, and most nearly negligible among troops safe in the United States. The A. E. F. was largely an army by compulsion. Logically, the tendency to desert would be greater among conscripted troops than among volunteers. These overseas troops, moreover, faced warfare deadlier than any the world had ever known; and the very sea which they must cross to reach the scene of conflict was full of peril. If ever cowardice and reluctance were to show themselves, it was when the American Army was departing for the unknown shores of France.

It is extraordinarily gratifying, then, to note what actually occurred. It may be years before the exact figures of American desertion in the World War will be compiled; but we can form an approximate estimate of it from the records of Camp Merritt. The camp handled in all about 50,000 casuals of the straggling class. Of these men, a great number were the merely unfortunate—soldiers left behind in hospitals to recover from accidents or illness, or men who had somehow lost themselves beside the way. The Camp Merritt figures show that about half the stragglers, or 25,000, possessed no military record papers and cards. We may assume that most men of this class had fallen behind by their own fault. The Army did not rate a man as a deserter unless he was absent without leave for more then ten days. Of the prisoners who passed through the Camp Merritt stockade, only about one-fifth were actual deserters. The others were men who had left their units without permission, and the great majority of them, after short absences, had returned voluntarily to the service. Of the more than 1,500,000 men who went to France *via* New York, only 5,000 remained behind and were apprehended as deserters.

This number compares favorably with the desertions from the comparative handful of American volunteer troops who fought in the Revolutionary War. There were more than

5,000 deserters from the Union Army in the *débâcle* which followed Bull Run. In fact, no American volunteer army of the past was nearly so steadfast in its fidelity and courage as the force which fought in France. Actual desertion, at least on this side of the ocean, was practically an invisible quantity. No doubt the superior training and discipline of our troops in 1918 kept desertion to a minimum, for this capital crime of the military service has always been most prevalent among green troops and troops held under loose restraint; yet something must be said for the spirit of the soldiers.

Of the prisoners at Camp Merritt, about four in ten were men who went absent without leave after their organizations reached the Port of Embarkation. They were the men who could not withstand the temptations of the city; men who allowed themselves to be persuaded to overstay their leaves, or who went on sprees and awoke to find themselves delinquent. On the theory that it is as well to be hanged for an old sheep as a lamb, many of these men continued A. W. O. L. until close to the time limit at which desertion begins and then returned voluntarily for whatever punishment they might receive. There was also an element in the service so ignorant as to believe that if they took French leave and were absent when their units sailed, they could then safely report themselves and be sent to prison on this side of the ocean, thus escaping foreign service altogether; and some of them put their theory into practice. They were unaware that deserters and men absent without leave were always forwarded to the A. E. F. to undergo punishment in France.

A good share of the prisoners in the stockade, however, were the weak and the unwise and the homesick. Upon learning that their organizations were scheduled to depart from the training camps for the port, numerous boys took the long chance of returning home without leave to say good-bye—most of them, no doubt, hoping and expecting to catch their units before they embarked. Some of those who went absent without leave at Camp Merritt eventually turned their misdemeanor into the crime of desertion by remaining away. Immediately after the armistice was declared, about 350 deserters appeared at Camp Merritt and gave themselves up.

Absenteeism at the port grew so serious in the summer of 1918 that the commanding general issued public appeals to the people of New York not to tempt soldiers to overstay leave, and to help men to return if they were so intoxicated that they did not know what they were about. On one occasion a unit of 1,800 troops, when it checked up on the pier, found 300 men missing. There were numerous instances of units sailing minus any number from 25 to 200 of the men whom they had brought to the port. The overpowering fascination of New York to thousands of boys who had never seen the city before was accountable for a large part of the absence without leave in that vicinity.

At first the Army forwarded prisoners to France, there to be tried by their own organizations. This plan turned out not feasible, and a permanent military court was set up in Camp Merritt to sit every day and dispose of cases. It had little trouble with men A. W. O. L. Most of these culprits came to camp voluntarily, with confessions on their lips. But desertion was a different matter. In time of war it was a capital crime, and a trial for desertion was a grave affair. The competent witnesses in a deserter's case—the comrades and officers of the man's own organization—were all on the other side of the Atlantic. When it was found impossible to convict men of desertion in the Camp Merritt court, the regulations were changed to send alleged deserters to the A. E. F. for trial.

When a man was brought to the Camp Merritt stockade, the officer on duty there issued to him a fatigue uniform, assigned him to quarters, and within twenty-four hours had him up for investigation before an officer who conducted a sort of preliminary court. This examination might acquit him or even release him with a warning, if the offense were trivial; or it might hold him for trial by court-martial. In the second event, he went before the court immediately. If found guilty, he was at once assigned to a skeleton company within the stockade. A prison casual company consisted of sixty-six men, all that could live in one barrack building. As soon as a prison company reached its full strength, it was sent under guard to the piers and embarked upon the first transport which had room for it.

The period between the sentencing of a man by the court-martial and his departure for Hoboken in an overseas prison company was not often longer than forty-eight hours. In that interval the camp made out his service records as best it could, issued to him a complete overseas equipment just as if he were a member of a regular organization, and restored to him any property which he had abandoned when he went A. W. O. L. He traveled to France as a prisoner and served his sentence there.

The stockade guard company was kept busy bringing in prisoners from all parts of the country. Its members traveled as far west as the Pacific coast. The actual coward, the soldier whose metal was dross, was extremely rare; yet there were a few whose moral nature was so warped that they preferred even suicide to a voyage across the ocean. Some faltering spirits simply could not stand the gaff. One prisoner who was being brought to Merritt under guard threw himself from the window of a swiftly moving train. The fall failed to kill him, and he got up and hanged himself to a telegraph pole with some wire which he found beside the track. Another who was marching at night with a company of prisoners from Camp Merritt to Alpine Landing took the frightful leap over the edge of the Palisades. In the darkness his deed was not witnessed, nor were the guards able to tell how he had escaped. Some days later, a gathering of buzzards caught the attention of the guardsmen. They gave search and learned the truth about his escape. Occasionally men would disappear after the ferry-boat left Alpine Landing and before it touched the pier in the lower river. These desperate individuals had jumped overboard. Some of them perhaps succeeded in swimming to shore; others never made it. Intermittently during the period of heaviest embarkation the watermen of New York fished the bodies of dead soldiers out of the river. A few soldiers, when the transports were moving from the piers down the bay to the rendezvous of the convoy, leaped to death in the deep water.

But for every such dark incident, there were a dozen of another sort—the stalwart youth condemned by some final inspection to remain behind because of alienage or physical disability, standing before the embarkation officer with unashamed tears flowing down his cheeks as he pleaded for the chance to share with his comrades the dangers ahead. This was a common sight at Hoboken in those dramatic days. The Government knew how rare were desertions and cowardice, how frequent the exhibition of valor; and it reposed a mighty faith in the moral stamina of the American Expeditionary Forces.

From An Official Motion Picture

CASUALS AT CAMP MERRITT RECEIVING EMBARKATION INSTRUCTIONS

From An Official Motion Picture

ENTRANCE TO STOCKADE, CAMP MERRITT

Photo by Signal Corps

LEVIATHAN LEAVING FOR FRANCE, AUGUST 3, 1918, WITH NEARLY 11,000 YANKEE TROOPS

Photo by Signal Corps

ON ONE OF *LEVIATHAN'S* DECKS, AUGUST 3, 1918

CHAPTER XVII

THE EMBARKATION SERVICE

THE New York Port of Embarkation antedated the Embarkation Service, of which it was an integral part during so much of the period of overseas travel. The Port of Embarkation, we have seen, was created as a military entity in July, 1917, a few days after the first convoy had departed for France; but it was not until August 8 that the Embarkation Service came into existence.

It may confuse the civilian reader to find the function of troop transportation carried on, as it was during our hostilities with Germany, by the General Staff. The layman thinks of a staff as an advisory and not as an executive and operating body. Such, in theory, is our General Staff—a deviser and creator of plans and policies, not the organ for executing them—and such it was in fact, before the declaration of war in 1917. The expansion of the American General Staff as an operating agency, its assumption of duties which logically and normally fell to the administrative branches of the Army—this is a significant chapter of American military history. Some day, no doubt, wise men will make studies of this growth and draw conclusions therefrom; but the province of this narrative is only to chronicle the facts of the expansion in so far as they pertain to the transportation of troops and supplies.

In theory, the General Staff in Washington is the Army's executive committee. It may be likened to the board of directors of a railroad. The directors are men of deep experience and skill. In their meetings they lay broad plans for the growth and efficient administration of the property. But it is not their function to carry out these policies in executive administration. That duty falls to the operating staff of the road—its line organization, to continue the military analogy—which consists of the railroad's president and his assistants, the general superintendent, the general freight agent, the general passenger agent, the general superintendent of rolling stock, the general superintendent of motive power, the general superintendent of maintenance of way, and others, each assistant being in actual executive control of some necessary branch of railroading. In the Army this organization is, or rather was, duplicated. The Commanding General in the field (or the Adjutant General in time of peace) acted as the railroad president does; the Chief of Staff was the chairman of the board of directors.

Up to the sixth day of April, 1917, the General Staff consisted of (1) a small conclave of officers on duty at the office of the Chief of Staff in the War Department, and of (2) a group of military academicians gathered together for research at the Army War College in Washington, and known collectively as the War Plans Division of the General Staff. The War Plans Division had its chief, a secretary, and certain committees, which might be likened to the several faculties of a university. Each committee conducted some special branch of investigation. There was, for instance, the Organization Committee, which, through its observers stationed with the armies of other nations and through its own studies of domestic conditions, from time to time recommended changes and improvements in the organization of the American Army. It was the Organization Committee which, in conference with the British and French officers sent to America with the first visiting missions, drew up for General Pershing the plan, hereinbefore mentioned, for the organization of an expeditionary division with enlarged companies and other innovations in harmony with the organization of the French and British armies.

Another committee at the War College was the Operations Committee, whose responsibility was to work out plans for our actual field operations, including the maneuvering and disposal of troops at the front and their transportation to the scene of action. In years past, this committee had built tentative plans for the manipulation of American forces against various specific but supposititious enemies, so that an unexpected emergency might not catch the United States strategically unprepared. The Operations Committee kept a vigilant eye on international politics; and any threat of possible future complications, no matter how remote, was sufficient to cause it to inaugurate studies and develop plans for meeting any conceivable situation. When the World War broke out and threatened more strongly month by month to involve America, this committee concentrated upon methods for the possible employment of our forces in Europe.

A third committee at the War College studied the multifarious problem of equipping our Army with clothing and ordnance and all the other supplies necessary to a great force. A fourth formulated systems for training soldiers. Out of this group came the projects for universal training considered by the Government before the outbreak of hostilities. A fifth committee procured military intelligence; it was this agency which employed the services of our military *attachés* in many lands.

When war was declared, this whole War Plans Division threw itself into the task of laying down a prospectus for raising, training, equipping, and transporting an army of forty-two divisions, with their necessary maintenance troops. Such a force would number, roughly, 1,600,000 men. It will be seen that this program was considerably in advance of any conceived at that time (the spring of 1917) by the line organization of the Army. In fact, General Pershing, in October, 1917, when he sent his first great requisition for troops, called for a force that should consist of but thirty divisions, with the appropriate army troops, corps troops, and troops for the line of communication. It was the province of the War Plans Division to step out in this fashion in advance of contemporary needs; to be ready with practical measures when the Government's power to organize troops had reached greater stages of development.

It was soon discovered that this closet organization, the War Plans Division, was too remote from the control of the Army to be effective; and little by little, at first by the inclusion of liaison officers within General Staff headquarters, these committees were brought out of the War College, until at length, vastly expanded in personnel and power, they became independent divisions of the General Staff itself. The Operations Committee, for instance, grew into the powerful Operations Division of the General Staff, an organization whose behests set in motion the machinery of the draft, built up the combat divisions at the training camps, called into being the bewildering numbers of special troop organizations, dictated the transportation of all units, and fixed priorities in overseas travel. The Operations Division also supervised the distribution of supplies. The Equipment Committee eventually united with the whole quartermaster branch of the Army to become the great Purchase, Storage, and Traffic Division of the General Staff, which by that time had actually assumed the duties of buying all army supplies (except technical equipment) and of transporting supplies and men.

By the close of 1917, the General Staff, without losing any of its advisory powers, had taken on vast responsibilities in the actual executive operation of many of the most vital functions of warfare. Herein our General Staff differed from the staffs of other modern armies, for they remained only advisory. It was natural, even inevitable, that our Staff should branch out in these administrative directions. It included within itself the very men who knew most about these subjects of transport and purchase. The expedient thing for the Staff to do was to assume direct charge of the great supply and traffic enterprises made necessary by the war, instead of creating new organizations for these purposes and instructing them afterward.

We are now in a position to understand the place of the Embarkation Service in this system. In early August, 1917, Major General Bliss was Acting Chief of Staff. His assistant was Major General Kernan. At this time the Staff was attempting, through the Operations Committee, merely to coördinate the movement of troops and supplies; that is, it had no operative powers in this quarter. Each military bureau still retained authority to transport its own troops and supplies as it chose, and each conducted its business with heedless disregard of its interference with the affairs of others. General Kernan felt that the War Department needed a central agency to coördinate the traffic, or at least that part of it which involved troops and supplies approaching the coast for transfer overseas; and in this judgment General Bliss concurred. And he made his concurrence effective by ordering the creation, within the General Staff, of a section known as the Embarkation Branch. At this time the Operations and Equipment Committees were staff branches; and so the Embarkation Branch was born on a parity with them. The blood brotherhood served our military transportation in good stead later on, when the success of the system depended crucially upon the perfection of the *rapprochement* between the Operations Division of the General Staff and the Embarkation Service, then a branch of the Purchase, Storage, and Traffic Division of the General Staff. The complete harmony established in these days of swaddling clothes continued to the end between the agency which ordered embarkation and the service which actually conducted it.

General Bliss made General Kernan the Chief of the Embarkation Branch. For his assistant, the Staff went over into one of the operating branches of the Army, the Quartermaster Department, and therefrom picked an officer who had had greater recent experience in the transportation of troops and supplies than any other man in the Army, Colonel Chauncey B. Baker. The administrative agency of military transportation prior to 1917 was the Transportation Division of the Quartermaster Department. Colonel Baker was chief of this division. His most practical experience in overseas transportation had been gained when the American military and naval forces occupied Vera Cruz in 1914. He had proceeded to Vera Cruz with the first ship and acted as base quartermaster there until our forces were withdrawn.

Still a third officer joined the new organization; a man whose personality was later to be so impressed upon our transportation history—for he was destined to become the chief figure in it—that it is of interest to examine his career in more detail.

When the United States declared war against Spain and the country was aflame with war enthusiasm, a young civil engineer named Frank T. Hines enlisted in the United States Volunteer Army as a private. He found a place in a battery of field artillery which was proceeding to the Philippine Islands. Out of its campaign there he emerged a few months later with a record for courage and ability and a commission as second lieutenant. One duty after another kept him in the uniform until, at the end of two years, he decided to make the profession of arms his own. He took up various technical branches of the service, working in the graduate army schools, until he had made himself an electrical and mechanical engineer, a captain in the Coast Artillery Corps, and a specialist in the location and equipment of coastal fortifications.

The Government of Greece, no doubt anticipating the clash of arms that was to resound throughout Europe, asked the American Government to lend it an officer to supervise the strengthening of the Hellenic coast defenses; and for this mission the War Department nominated Captain Hines. He was in Greece when Germany began her invasion of Belgium. He stuck to his work until, one day, he discovered that commercial travel at sea had ceased. His job had come automatically to an end. The last passenger vessel for the United States had departed; the steamship companies could promise no other. On board one of the wallowing tubs of the Ægean, Captain Hines crossed to Brindisi, on the Italian boot-heel, and thence made his way to Naples, where an important duty faced him.

Europe was crowded with American refugees, many of whom were women and children. They were stranded, practically without funds and without hope of getting home. Their immediate resources were generally in the form of letters of credit—many of them letters of credit issued by German steamship companies. In the financial paralysis of those first astonishing weeks of war, all instruments of international credit, whether issued by German institutions or not, had collapsed. An order for thousands of dollars was not good for a night's lodging. The State Department stepped in at this juncture to rescue the unhappy American victims of the cataclysm. The Secretary of State cabled to our ambassadors and ministers in Europe to manage the relief work in their respective jurisdictions; and the advices empowered them to command the services of any American army officers who were traveling in Europe. Mr. Thomas Nelson Page, the American ambassador at Rome, straightway discovered Captain Hines and placed him in charge of the work in Italy.

There were some 3,100 American tourists caught in Italy, and they were in a wretched plight. Most of them were entirely without money, living at inns and hotels solely on their verbal assurances that everything must come out all right. Some of the women were half hysterical with fear, and even the more courageous and hopeful found the outlook anything but bright.

It was impossible, in this dangerous time, for the American Government to induce foreign steamship companies to risk operating their vessels across the Atlantic. The naval situation was still uncertain; German raiding ships, and even German ships of war, were abroad on the seas. Moreover, the chief transatlantic lines were owned by the very nations which were at war, whose first steps had been to commandeer all tonnage that could be used in the transportation of troops and supplies. America then conceived the plan of building temporary passenger accommodations on several of our transports and naval cruisers and sending these vessels abroad with gold for the financial relief of the stranded and with quarters for their return passage. The Government actually carried out this plan, sending, among other ships, the cruiser *Tennessee* to Europe. By the use of her own ships and by crowding full the few

British and French liners which presently ventured to resume operations, America managed in four or five months to gather home her hapless nationals.

While the improvised American relief organizations in other countries were beseeching the home Government for passenger vessels, Captain Hines, in Italy, was ready with a plan which made no demand upon outside help. Searching the Italian harbors, he had found in Naples and Genoa four large, commodious passenger vessels, the property of an Italian trading concern. They were immigrant ships in the South American trade. Not even a sanguine person could have called them palatial in their appointments, for their accommodations were steerage throughout, except for a few second-class cabins for the wealthier colonists. But Captain Hines saw that he could quickly make the ships serviceable for the emergency work in hand. Aided by the backing of the American ambassador, he struck a bargain with the company, got the four ships, tore out their steerage accommodations, and built them full of temporary staterooms. While this work was going on, the officer, assisted by the American consul at Naples, arranged with the Italian banks a credit which gave him money to settle the bills of the refugees and buy their tickets home. Six weeks later the last of the 3,100 Americans sailed from Italy for New York, and in comfortable accommodations, too. At this early date not one of the American vessels then being outfitted for similar work had yet left the home shipyards. Our refugees in Italy were all in the bosoms of their families relating their adventures while the relief of those in other countries was still a vexing problem.

This episode proved to be the turning point in Captain Hines's career, for it marked him as an executive who could step out firmly on unbeaten paths. He was acting as a member of a Coast Artillery board at Fortress Monroe, Virginia, when the United States declared war against Germany. The Army Staff at once reached out for him and assigned him to the Equipment Committee at the Army War College; and in August, when the Staff created its Embarkation Branch, General Bliss made Captain Hines chief of staff in the new organization, in which capacity he acted as executive officer in the first attempt at centralized control of embarkation.

The Embarkation Branch began its existence with the three officers mentioned, General Kernan, Colonel Baker, and Captain Hines, and with a single civilian clerk. The first thing that attracted the attention of the junior officer of this group was the slowness of the War Department in securing vessels for the transatlantic shipment of the great supplies of materials for dock-building which the Engineer Corps was accumulating at Jacksonville and Fernandino, Florida. In many ways this was the most vital cargo that could cross the ocean at that time, for until we had provided adequate port facilities in France the growth of the A. E. F. must inevitably be slow. On the advice of Captain Hines, Colonel Baker went to the United States Shipping Board and secured its promise to allocate a greater amount of tonnage to the engineering cargo.

In August, 1917, the competitive system of military transportation was in full swing. Every production bureau in the War Department was trying to be the first to deposit its supplies in France, and every special corps vied with the others to get its troops shipped overseas. By September the Embarkation Branch had established a port-release system of a sort designed to control the flow of export army freight tonnage. Yet the Embarkation Branch was still merely an advisory body, still in essence a staff organization; and its advisory system of release was not sufficient to control the congestion that steadily grew at the seaboard. Captain Hines wrote for the Chief of Staff a memorandum to that effect, in which he stated that the embryonic service, though it might exercise some degree of regulation of purely war department freight, had no control over domestic shipments to the New York Quartermaster Depot, and none over the traffic in munitions and supplies obtained in the United States by the Allies. War department export freight was as yet only a small percentage of the total volume of traffic moving into New York. No matter how much this advisory body could systematize the military freight movement, it could not forestall the general traffic confusion then imminent.

The aftermath of this memorandum was the creation of the so-called Coördination Committee. This was a committee made up of representatives of the railroads interested in traffic at New York, together with the representatives of the British and French Missions, the War Department itself, the Navy, the Shipping Board, and what was then the germ of the future United States Food Administration. This committee met once a week with the American Railway Association's Committee of Five in Washington. These meetings represented the Government's first attempt to regulate all traffic. By this time every official agency, including the railroads and the representatives of the Allies, had established its own independent release system at the port of New York. The various releases were brought each week before the Coördination Committee, which attempted to give precedence to them in proper order.

The intention of the committee was good, but its authority was weak. Each representative on the committee possessed executive control of shipments for his own department; but the committee itself lacked that overlordship which alone could have made it an effective power. A few weeks later it gave way to the famous Priorities Committee of the Council of National Defense, headed by Judge Lovett, the eminent railroad man. This committee, which acted in coördination with the American Railway Association, possessed in fact, if not by actual law, the authority to control freight movements by rail. It did succeed to some extent in gaining upon the congestion at the port, but it was deficient in the summary power to deal with shipping at its points of origin—which power, as the event proved, was the *sine qua non* of effective traffic control.

When the severe winter of 1917-1918 closed in, it brought a traffic paralysis utterly appalling to those in a position to observe its effect upon overseas shipping. The Government promptly seized the railroads—an act which, as we have related, gave the War Department its opportunity to establish the Inland Traffic Service, one of the great agencies which composed the Division of Purchase, Storage, and Traffic. Shortly before this event, even, it had become evident that firm-handed centralized control was necessary, not only in the shipment of supplies, but equally in the whole direction of the supply enterprise; and the result was the immediate creation of the Purchase, Storage, and Traffic Division of the General Staff. The direction of this centralized supply agency demanded an executive of qualifications which few men possessed—courage in shouldering responsibility and a willingness to cut corners and drive through obstacles to any desired end. For this post the War Department selected Major General George W. Goethals, who had built the Panama Canal, the largest work ever undertaken by the Government before 1917. Prior to this appointment, General Goethals had been Acting Quartermaster General. He brought into the Division of Purchase, Storage, and Traffic the new Inland Traffic Service, and took over also the Embarkation Branch of the Staff, raising it to the status of an independent service. Thus, in addition to his other powers, General Goethals, with both the Inland Traffic Service (rail) and the Embarkation Service (port and ocean) under his jurisdiction, became the master of military transportation. Of the two traffic services, the Embarkation Service was the greater, for the Inland Traffic Service maintained no control over any activity of the Embarkation Service, whereas the Embarkation Service was dictator to the Inland Traffic Service in one of its most important operations—the transportation of troops and supplies to the seaports.

But there was one significant development which we have still to mention. In the autumn of 1917 General Kernan was assigned to duty in France. Colonel Baker, now wearing the stars of a brigadier general, was made chief of the Embarkation Branch of the Staff. Captain Hines, who had become Major Hines, was appointed assistant chief. General Goethals found this organization when he took charge. It did not take the canal-builder long to discover that the embarkation control had not been functioning as it should. General Goethals attributed its ineffectiveness to the lack of a practical shipping man at the head. Therefore he transferred General Baker to another branch of the military service and, during the interval of several weeks in which he was hunting for the practical shipping man to be the civilian head of the Embarkation Service, left Major Hines in charge as acting chief. Major Hines seized the opportunity to upset completely the organization of the Embarkation Service as created by General Baker and to remodel it according to his own ideas.

In January, 1918, General Goethals found his man. He was Mr. J. L. Lilly, of the ocean shipping concern of Norton, Lilly & Company. General Goethals made Mr. Lilly Chief of Embarkation, but retained Major Hines, who by this time was a lieutenant colonel, as assistant chief. Mr. Lilly accepted the appointment, took his place in the Embarkation Service, signed the official communications, and studied the organization of the Embarkation Service to analyze its weakness and its strength. Five days later he went to General Goethals to express his conviction that he could best serve the Government by acting as a subordinate in the Embarkation Service. He told General Goethals that even his long experience could contribute no suggestion for improving the system which Lieutenant Colonel Hines had built, and he strongly urged that Hines be made Chief of Embarkation and given a free hand in the direction of the service.

General Goethals acted at once upon this advice. He made Lieutenant Colonel Hines Chief of Embarkation, and Mr. Lilly went to the Port at New York, there to serve in several important capacities throughout the rest of the embarkation period. A few weeks later Lieutenant Colonel Hines was promoted to a full colonelcy, and in April, 1918, he was made a brigadier general, having advanced to that eminence from the rank of captain within eight months. General Hines continued to serve as Chief of Embarkation until after the armistice. Then occurred a development contemplated before the cessation of hostilities: one which, even had the war continued, would undoubtedly have come to pass as it did. In December, 1918, the Inland Traffic Service and the Embarkation Service were merged in a single organization known as the Transportation Service, and General Hines took command as chief of it.

Second in importance only to the actual maneuvering of the American Expeditionary Forces in the field was the transportation of the Army. Indeed, it will probably always remain a subject of at least academic dispute which was the greater triumph of American prowess—the transportation of the two million men to France in little more than a year, or the maneuvering of that force in France. We are safe in saying that these enterprises were easily the two most momentous episodes in the history of our Army in the World War.

The abrupt rise of a relatively obscure captain of the Coast Artillery to the command of this vital service was only another instance of the devastation wrought by war within the established military bureaus. It is no derogation from the inherent ability of many officers who for years prior to 1917 had held some of the chief posts in the military service, to emphasize the bare fact that the great test of the war unseated those officers and, often, raised to their places men theretofore comparatively unknown. Those of extended experience in army bureaus, no matter how great their original stock of initiative or native ability, become chained by long association to regulation and precedent. There is no help for it; it is so in every army. The system will take the most individual of men and slowly but surely mold him into the common form. When the great emergency comes, such men often find themselves too timorous, too fearful of transcending custom, to make executives of the best type. When a nation is committed to a struggle for existence, only a man impatient of hampering custom is likely to carry a great project through to success. Such a man was General Goethals, and such was General Hines. The very freshness of these men in their work, their lack of previous intimate contact with the red tape and machinery of the war bureaus, fused with their native ability, judgment, and determination to make them successful executives. They were bold in assuming responsibilities, willing to strike out in new directions without driving their superiors to distraction by continual requests for authority to act.

General Hines's first act as Chief of Embarkation was typical of his whole administration. Up to this time the Embarkation Service had been going to the Shipping Board hat in hand and deferentially requesting its quota of ships for war department freight. General Hines did not request, but—none the less politely, of course—demanded all the vessels the Shipping Board possessed. To be sure, he did not get them—he scarcely expected that, for other important functions of the Government had to be served. There had to be ships in the Chilean nitrate trade, ships to bring the indispensable manganese ore from Brazil and Cuba, ships in the service of the Food Administration carrying relief to Belgium and other parts of stricken Europe, ships placed at the service of the Swiss to keep them content in their neutrality. But by demanding all the tonnage in sight, General Hines acquired a greater portion of it than he could have got by giving obsequious consideration to these other needs.

When he took office there was still a great accumulation of engineering materials at the seaports—materials for use in construction of docks at Bordeaux, St. Nazaire, and other French ports assigned to the A. E. F. The additional shipping placed at General Hines's disposal soon cleared away this accumulation. Had the Engineer Corps been able to prosecute its work on the French ports at the rate expected in the summer of 1917, the Embarkation Service would have scored a failure to deliver the material. So narrow was the margin of success that abilities of even the first brilliancy needed the intervention of pure good fortune.

Meanwhile certain activities within the Embarkation Service had now expanded to great importance. The Embarkation Branch at first kept manifests of export freight which showed only the general content of packages shipped abroad. Soon there began to come from General Pershing cabled demands for specific information about the shipment of various consignments. The Embarkation Branch could not answer such questions, nor could it help the A. E. F. trace any missing supplies, because its records did not contain the necessary information. To supply this defect the Branch set up its cargo section. The cargo section kept detailed records of the contents of every package sent to the A. E. F. as well as of the quantities and kinds of all bulk freight. Under the direction of Major Morse, a former official of the Pennsylvania Railroad, this work grew until it required the services of 180 clerks. Eventually the section kept track of all important supplies from the moment they were packed in the shipping rooms of American factories.

A similar system was adopted in the movement of troops. Whenever General Pershing cabled for troop units, of whatever branch of the service, copies of his cablegrams went to the Embarkation Service, where the proper employees placed in the record index a card for each unit requested. When the first entry was made on a card, there was as yet no such unit in existence: there was only a requisition for it. Weeks or months later, such a unit would be organized. The Operations Division of the General Staff would place it on the priority list for overseas transport and send notice of this action to the Embarkation Service, which thereupon made the suitable entry on the card. Presently the port machine worked down through the priority list until it neared this unit. Thereupon the Embarkation Service issued a release for the travel of the unit to the port and, through the Adjutant General and the regular military channel of communications, ordered the unit to proceed to the port; but—and this is most important—not before the troop commander had communicated with the Commander of the Port and from him received specific orders for the travel.

Right here is the secret of the successful embarkation of the two million. The Embarkation Service did not attempt a centralized control of overseas travel. If the Washington headquarters had ordered the unit to the port and simultaneously had ordered the Port Commander to receive the unit and embark it upon a transport, the whole overseas travel system would soon have come to grief. Washington was too far removed from that hour-by-hour intimacy with conditions at waterside to attempt to manage the details of embarkation. Instead, it created a joint responsibility for the travel of a unit to France and placed it partly upon the unit's own commander and partly upon the Commander of the Port. It was not unduly difficult for those two to arrange the travel without interference to the travel of other units.

A similar system was applied to the overseas transportation of individual officers. As soon as a requisition from the A. E. F. called for an officer of certain qualifications, a numbered, but nameless, card was started in the Embarkation Service's index. Later, when an officer was appointed to the assignment, his name was entered on the card, and in due time the Service made a place for him on a transport.

The activities of the Embarkation Service ramified, we have seen, in many directions. As has been stated, it exercised command over the New York Port of Embarkation. Later it established expeditionary depots at Baltimore, Philadelphia, and Boston. These it joined to the New York Port of Embarkation as subsidiary ports. It created one other independent port—that at Newport News. Troops and cargoes were embarked and shipped from Portland, Maine, Montreal, Quebec, and even from Halifax, Nova Scotia; but these were emergency ports, used only occasionally; and whenever they were in such use they were under the command of the New York Port of Embarkation. Newport News possessed no subsidiary ports. The Embarkation Service also established its famous courier service for the speedy transmission of confidential or other important communications back and forth between the A. E. F. and Washington; and it conducted other enterprises related to transportation.

But fundamentally the most important activity of the Embarkation Service was its incessant and ineluctable struggle for ocean tonnage adequate to the overseas movement. The overwhelming disaster contingent on mismanagement of the tonnage situation was a reality which earlier administrations of the Service had never fully grasped. We were forcing the overseas embarkation of troops to the extreme limit, a limit which, before it was reached, the military experts of the world had deemed impossible of attainment. The haunting fear of the embarkation authorities was that they might dispatch to France more troops than they could find the tonnage to maintain. If that situation ever arose, it meant disaster to the American arms, if not to the Allied cause.

The first great triumph of the Embarkation Service in the direction of procuring tonnage was its acquisition of the British troopships in the spring of 1918—an achievement most brilliantly registered in the stupendous figures of overseas sailings in the memorable spring and summer of that year. With embarkation proceeding at such a rate, the burden of finding the necessary supporting tonnage was more crushing than ever. The Service and its director were the decisive factor in the acquisition of the Dutch tonnage and, in the late summer of 1918, of the British cargo tonnage—both great episodes in the enterprise. Moreover, the exhaustive studies of ocean traffic which issued from the brain of the Chief of Embarkation were among the notable documents produced in our Army during the World War. It was upon them, in a real and far-reaching sense, that the War Department based its whole program of operation.

CHAPTER XVIII

ORDERS AND ITEM NUMBERS

AT Fort Benjamin Harrison in Indiana were three battalions of the 34th Regiment of Engineers, operating a military supply train and a mobile repair shop. They numbered some 50 officers and 2,400 enlisted men, and every one of them was impatiently awaiting the order that would send them all overseas.

Expectancy had been acute since the 14th of May, 1918, when the battalions had received from the Commander of the Port of Embarkation at Hoboken, New Jersey, the well-known and coveted form letter beginning: "Instructions have been received from the War Department that your organization has been designated for service abroad. . . . " The letter went on to tell them what preparations they were to make for sailing. Except for its address, it was identical with the letters sent to all military units about to sail for France—the first harbinger of their embarkation.

But it was now mid-July. Weeks ago the three battalions had followed the mimeographed instructions of the letter; weeks ago they had secured all of the prescribed paraphernalia that they could lay hands on. Two months had passed and brought no order to proceed to the port. The newspapers were black with headlines flaunting tremendous doings in France. The Germans were at the height of their successes; the Channel ports were threatened; Paris was under bombardment; and American soldiers by hundreds of thousands were in the thick of the fighting, writing their imperishable record on the pages of history. And out in this peaceful interior post, surrounded by a smiling Hoosier landscape that dreamed beneath a summer sun, lay these engineer troops, eating their hearts out with chagrin because of an order that never came.

Washington had not forgotten them. Troops in camp could scarcely be expected to know about the almost daily military crises that upset the orderly arrangements for travel during those momentous weeks. Plan as they might in advance, the military heads in America could not begin to anticipate the needs of the A. E. F. as those needs were modified and controlled by events at the front. And, although organizations might be on the list for early sailing, their departure was often postponed in order that the Transportation Service might hurry to France, out of turn, specialized troops of one sort or another demanded forthwith by the A. E. F. cables.

The day came when Washington telegraphed the following message:

Adjutant General's Office
July 20, 1918

To be sent in broken code.

Commanding General,
Central Department,
Chicago, Illinois.

Send regiment headquarters, First, Second, and Fourth Battalions of 34th Engineers (Supply and Shop), pertaining to Item E-403 and E-404, 3d Phase, consisting of 56 officers and 2,410 enlisted men, now at Fort Benjamin Harrison, Indiana, to Port of Embarkation, Hoboken, N. J., after arranging time of arrival and other details directly with Commander of the Port. Do not entrain troops until Commander of Port advises you that he is ready to receive them. Have inspection made to determine if organizations and individuals are properly supplied with serviceable authorized clothing, equipment, and medical supplies, reporting result by telegram. Anything found lacking to be reported in detail. Leave enemy aliens behind. If any so left, report number.

McCain.

This was a typical overseas order; except for the substitution of other names and addresses, it went to thousands of expeditionary organizations. It reached its destination "through military channels," which in this instance happened to be the Commander of the Central Department at Chicago. To find whether the commanders of the battalion hastened to comply with the orders and to ask Hoboken to arrange the details of their arrival at port, we turn to the next important document in the overseas file of the 34th Regiment, a telegram sent six days later by Major General David S. Shanks, Commander of the Port of Embarkation, Hoboken, New Jersey.

(*Translation*) SJC/MS

CONFIDENTIAL

July 26, 1918

Commanding General, Central Department,
Chicago, Illinois.

T-125 Request you to send to Camp Upton, Long Island, New York, to arrive not earlier than noon August 6 and not later than noon August 8: Regimental Headquarters, 1st, 2d, and 4th Battalions, 34th Engineers, 56 officers and 2,410 enlisted men, Item No. E-403 and E-404, 3d Phase, now at Fort Benjamin Harrison.

Leave at station all animals, ambulances, combat wagons, medical carts, ration carts, water carts, escort wagons, combat carts, spring wagons, passenger automobiles, motorcycles, cargo trucks, ammunition trucks, and tank trucks.

All other vehicles not mentioned above to be shipped as freight. Ship freight to General Superintendent, Army Transport Service, New York City, for lighterage on separate bill of lading and loaded in separate cars. Motor vehicles must be on second separate bill of lading and loaded in separate cars. Advise accurately weight and cubical measurement of freight, and number and make of all vehicles.

Troops to take with them to Camp Upton field ranges, field desks, authorized typewriters, office records, individual equipment of officers and enlisted men as outlined in Circular, War Department, July 11, 1918. All officers should be familiar with the various points of shipment of freight, vehicles, and baggage as above indicated.

Personal records including qualifications and locator cards should accompany troops.

Consult with representative of United States Railroad Administration regarding all details of train movements and conform to schedule as arranged with him. Telegraph Commanding General, Camp Upton, in advance time of arrival, names of organizations, number of men in each section, and list of shortages. Please acknowledge.

SHANKS.

Copies—

2 to Representative of U. S. R. R. Administration.
1 to Personnel Adjutant.
1 to General Superintendent, Army Transport Service.
1 to Shipping Control.
1 to Equipment Officer.

1 to Commanding Officer of Camp.
1 to Inspector of Camp.
1 to Inland Transportation, Washington.

In the sequence of orders and other communications which passed during the progress of an organization of troops from the training camp to its quarters on shipboard, one can trace clearly the process of embarkation. Each departing unit left in the archives of the Port the documentary history of its sailing. These paper records are essentially identical; to set down one of them is to represent all. In tracing the movement of the 34th Engineers, however, we have passed by one important paper—the release for the travel, emanating from the Embarkation Service. The Adjutant General could not act until he had this release before him. Each day the Adjutant General received from the Embarkation Service a number of releases for overseas travel, and his telegrams ordering the troops to the port were the result. The Embarkation Service was in full control of the troop travel to the ports, and only the army punctilio which required that troops receive operation orders from the Command of the Line, and not from the General Staff, kept the embarkation authorities from communicating directly with the troops.

The embarkation release for the 34th Engineers was as follows:

CONFIDENTIAL

July 19, 1918

MEMORANDUM FOR THE ADJUTANT GENERAL OF THE ARMY 541.1

SUBJECT: *Transfer of the following troops to Port of Embarkation, Hoboken, N. J.: Regimental Headquarters, 1st, 2d, and 4th Bns., 34th Engineers, Ft. Benj. Harrison, Ind.*

The Secretary of War directs that instructions be issued substantially as follows:

1. Direct Commanding General, Central Department, confidentially by wire to send the following troops to the Port of Embarkation, Hoboken, N. J., after arranging time of arrival and other details directly with the Commanding General of the Port:

Regimental headquarters, 1st, 2d, and 4th Battalions of the 34th

Engineers (Supply and Shop), Item E-403-404, 3d Phase, consisting of 56 officers and 2,410 enlisted men, now at Fort Benjamin Harrison, Ind.

2. Advise Commanding General, Port of Embarkation, Hoboken, N. J., of the above action.

By authority of the Director of Purchase, Storage, and Traffic.

FRANK T. HINES,
Brigadier-General, G. S., N. A.,
Chief of Embarkation.

RHJ/JIM

Copy to Hoboken.
Copy to Colonel McAndrews.
Copy to Inland Traffic.

The reader will have noted on the faces of some of these communications the arrangement for a considerable distribution of copies of them. This system was the device which made the work of embarkation almost automatic. One copy of General Hines's memorandum went to Hoboken and prepared the Port for the impending travel of the 34th Engineers. Having this notice, the port officers would look over their embarkation camp facilities and their transport accommodations for the near future, so as to be able to arrange for the reception and embarkation of the engineer regiment as soon as they received the organization's request for travel directions. Colonel Joseph R. McAndrews was the executive officer of the Operations Branch of the General Staff, the primary executive arranging the priorities for sailing. His copy of General Hines's memorandum showed him the progress being made in embarkation. The copy to the Inland Traffic Service was notification of the impending railroad travel of the 34th Engineers. The commander of the regiment would soon request the Fort Benjamin Harrison general agent of the troop-movement office to supply trains and schedules for the organization. The general agent, in turn, would relay this request to the troop-movement office in Washington, which, having received several days earlier a copy of General Hines's memorandum, would have the train equipment ready.

In the same way General Shanks, the commander at

Hoboken, directed a numerous distribution of copies of his orders to the 34th Engineers, notifying simultaneously the port representatives of the troop-movement office, the Army Transport Service, the Shipping Control Committee (which arranged for the reception of the regiment's freight), the commander of Camp Upton, who was to entertain the regiment, and numerous other port officers to whom the arrival of the unit meant work; sending one copy as well to the Inland Traffic Service in Washington, as a further precautionary guarantee that the train equipment would be ready.

General Shanks directed the regiment to arrive at the port between noon of August 6 and noon of August 8, 1918. Each embarkation camp sent daily to Hoboken a report of troops in camp awaiting ocean passage. In the file at Hoboken was a card which read as follows:

HEADQUARTERS, CAMP UPTON, L. I.

August 9, 1918

DAILY REPORT OF TROOPS AT CAMP UPTON AWAITING PASSAGE

Organization	*Officers*	*Enlisted Men*	*Equipped*	*Inspected*	*Date of Arrival*	*A. W. O. L.*	*Remarks*
Base Hospital No. 62	30	200	No	No	7/1	2	
Base Hospital Casuals		69					
88th Division, 338th Mach. Gn. Bn.	23	730	No	Yes	8/9		
Division Hdqrs. Bn.*	27		No	No	8/8		
Hdqrs. Troop	3	120	No	No	8/8		
Hdqrs. Detach.		99	No	No	8/8		
350th Infantry	78	3,382	No	No	8/8		
34th Engineers except 3d Bn.	35	2,403	No	No	8/8	3	

*3 British officers, 2 French officers, 5 Army Field Clerks not included

R. V. HISCOE,

Major Infantry, N. A., Adjutant.

The final item shows that the 34th Engineers, except the 3d battalion (which had not been ordered to port), arrived at Camp Upton on August 8, as scheduled, bringing 35 officers and 2,403 enlisted men, a strength somewhat shorter than had been indicated in the preceding records. On August 9 the camp organization had not yet inspected the regiment. Three soldiers had gone absent without leave.

Next day the embarkation machine began to draw the 34th Engineers into its hopper. The port authorities sent to Camp Upton a notification as follows:

HEADQUARTERS, PORT OF EMBARKATION
HOBOKEN, NEW JERSEY

SERIAL NO.
August 10, 1918.

TENTATIVE ASSIGNMENT

Transport No. 642 (England)
Pier No. 61, North River, New York City,

August 15, 1918

	Officers	*Men*	*Item No.*
34th Engineers Camp Upton, Long Island	35	2,403	Phase 3-E-403 E-404
Grand Total	35	2,403	
Capacity	100	2,303	

By authority of the General Superintendent, A. T. S.

C. E. HOOPER,
Captain, Q. M. R. C.

"Transport No. 642 (England)" was the British liner *Euripides*. Transports, both American and foreign, were called by number and not by name during the war. The assignment of the 34th to the *Euripides* was made tentatively because the embarkation officers never knew until the hour of sailing the exact space to be available on a ship; and unless all three battalions could get on the *Euripides*, not one could be permitted to embark on her. Military regulations forbade the splitting up of units in overseas transit. Note that the capacity of "Transport 642" lacked 100 berths of being great enough to accommodate the 34th Engineers. Excess assigning of this sort was customary. The port officers knew by long experience that seldom did an embarking organization come up to the piers at its full scheduled strength. Better to risk crowding the men a bit on board ship than to match accommodations in advance exactly to organization strength and then see a ship go out with empty berths.

On the same day, August 10, the dispatch office at Hoboken sent to the Commander of Camp Upton the following instructions:

L-451 August 10, 1918

FROM: Assistant Port Adjutant

TO: Commanding General, Camp Upton, Long Island, N. Y.

SUBJECT: Overseas Transportation

1. Transportation available on Thursday, August 15, 1918, for the following organizations:

On ship No. 557:

51st Telegraph Battalion, 9 officers, 206 men, Item S-103, 4th Phase; Field Hospital and Ambulance Company No. 39, 11 officers, 204 men, Item M-201, 4th Phase. These organizations should arrive at Pier No. 53, North River, New York City, N. Y., on Thursday, August 15, 1918.

2. On ship No. 556:

3d Battalion and Cos. G & H, 350th Infantry, 38 officers, 1,400 men.
Attached medical personnel, 5 officers, 24 men.
338th Machine Gun Battalion, 23 officers, 720 men.
These organizations should arrive at Pier No. 53, North River, New York City, N. Y., on Thursday, August 15, 1918.

3. On ship No. 642:

34th Engineers, 35 officers, 2,403 men, Items E-403, E-404, 3d Phase. This organization should arrive at Pier No. 61, North River, New York City, N. Y., on Thursday, August 15, 1918.

4. The following advance parties will report to Col. G. N. McManus, C. A. C., at Pier No. 1, Port of Embarkation, Hoboken, N. J., 9 A.M. Wednesday, August 14, 1918, prepared to go aboard ship and

remain there. One member of each detachment should be a stenographer, who should bring along a typewriter.

5. For ship No. 557.

Commanding Officer, Adjutant, Senior Medical Officer, and 3 enlisted men, from 51st Telegraphic Battalion.

6. For ship No. 556.

Commanding Officer, Adjutant, Senior Medical Officer, and 3 enlisted men, from 338th Machine Gun Battalion.

7. For ship No. 642.

Commanding Officer, Adjutant, Senior Medical Officer, and 3 enlisted men, from 34th Engineers.

8. Troops to take with them light and heavy baggage, personal equipment of officers and enlisted men, and all office records.

9. Triplicate lists giving names of officers, noncommissioned officers above Grade No. 17, and total number of other enlisted men, should be sent to these headquarters at the earliest possible moment.

10. In making routing consult with representatives of the U. S. Railroad Administration regarding details of train movements, and hours of departure and arrival.

Please acknowledge receipt by telegraph as soon as received.

F. F. Roy,
Captain, A. G., N. A.

Copies—
2 to C. G. Camp Upton
2 to U. S. R. R. Administration P. of E.
1 to Personnel Adj.
1 to Transport Q. M.
1 to Equipment Officer
1 to Director of Shipping
1 to Assistant to C. G.
1 to Dispatch Off.

Paragraphs 3 and 7 interest us. They specify transport, pier, and date of embarkation for the 34th Engineers and arrange for the customary advance party to go on board twenty-four hours ahead of the others, to be instructed in the ship routine and be ready to receive and settle in its quarters the rest of the regiment when it came up the gangplanks the next day. Pier No. 61 belonged to one of the British lines; it was located on the New York side of the North River.

This completes the historical file of the sailing of the 34th Engineers. After an organization had sailed, however, the Port made for its own records an index card giving tersely the essential facts and dates in the progress of the unit from training camp to transport. The following is the text on the card showing this record for the regimental headquarters and first battalion of the 34th Engineers:

FILE *508* ORGANIZATION *Regt. Hdqrs. & 34th Engrs. 1st Btn.*

OFFICERS *25* ENL. ABOVE GRADE 17 OTHER ENLISTED *794*

STATION *Ft. Benjamin Harrison, Ind.* CIVILIAN EMPLOYEES

ORDERED TO *Camp Upton* DATE *Between 8/6 & 8/8*

PASSENGER LISTS & EMBARKATION REGULATIONS SENT *5/15.* RECD. *7/6*

RELEASE *Par. 8 Sh. Sch. No. 1, 3d Phase; Hines 7/19*

AUTHORITY *A. G. 7/20 to C. G. Cent. Dept.*

ORDERED *Pier 61 N. R. 8/15*

ASSIGNMENT *Euripides* PIER *61*

DATE OF SAILING *Aug. 16, 1918*

ORDER Q. M. PORT OF EMBARKATION NURSES CIVILIAN EMPLOYEES

REMARKS *E-403 E-404 (Item No.)*

A perusal of this card gives one more than a little insight into the system of embarkation. The card is plain enough to anyone as far down as the entry beginning "Release," which is followed by certain cryptic abbreviations translatable into "Paragraph 8, Shipping Schedule No. 1, 3d Phase; Brigadier General Hines, July 19, 1918." This entry has reference partly to General Hines's memorandum releasing the unit for travel to the port, and further to what was, during the period of hostilities, one of the most secret of all documents in the possession of the War Department—Shipping Schedule No. 1.

Shipping Schedule No. 1 was compiled by General Pershing and his aides in the early autumn of 1917. It was dated October 7, 1917. The schedule was a product of the study which General Pershing and his officers made in France between the time of their arrival in early June and the date of the document. It was the architect's plan on which the A. E. F. was to be erected, and at the same time it was a set of instructions to the military organization at home as to the order in which troops should be sent to France. It provided for the formation of the A. E. F. up to and including its sixth army corps. The completion of the schedule would put in France thirty divisions of the line and their necessary maintenance and auxiliary troops in the Services of Supply and the line of communications.

The schedule was divided into phases, each phase providing for the shipment of a single army corps with its necessary maintenance organization. General Pershing's headquarters, uninformed of the individual identity of the units then being recruited and trained in the United States, made no attempt in Shipping Schedule No. 1 to specify organizations by name, but called for troops by general designations according to the sorts of units desired. The Operations Division of the General Staff on this side took the general requisition and made it a specific one by assigning to places in the schedule the old and new organizations that fitted the requirements. Thus, Paragraph 8 of the 3d Phase of that schedule, compiled in remote France far back in October, 1917, called for three battalions of Engineers to operate a supply train and a repair shop in the A. E. F.'s line of communications. By spring of 1918 there was such an organization in existence—the 34th Engineers—three of whose four battalions did operate a supply train and a shop. The Operations Division selected the 34th to fill the generic requisition of Paragraph 8 of the shipping schedule; and when the Embarkation Service had worked down to this point in the schedule, the 34th Engineers received notice to proceed to France.

The rest of the card is plain, except the final entry. In the space given to "Remarks" we find a mysterious label—a so-called "item number," or rather two item numbers—assigned to this unit. The numbers are E-403 and E-404, Phase 3. Doubtless the reader has noticed that these cabalistic numerals have appeared on all the communications reprinted above. They were written in the letter sent to the 34th Engineers by the Port of Embarkation back in May; in fact, this letter bore to the regiment its first notification of the item numbers assigned to it. The numerals appeared next in General Hines's memorandum to the Adjutant General. They were repeated in the telegram of the Adjutant General ordering the travel, and reiterated in the telegram from the Commander of the Port detailing the travel arrangements. They were carried in the assignment of the organization to its transport, repeated once more in the instructions from the dispatch office notifying the regiment when it should arrive at the pier, set down at the head of the passenger lists, and, finally, written into the card record of the unit's embarkation. Such unvarying iteration must signify that these numbers were important; and so they were—as important as anything in the whole transportation system. Item numbers were the means adopted by the A. E. F. to identify its component parts and to prevent them from getting lost in transit—an easier mischance than one might suppose.

To understand item numbers, we must revert again to the A. E. F.'s plan for its own creation, its Shipping Schedule. The construction of such a force is like what that of a modern skyscraper would be if the skyscraper were built up floor by floor. In the first place, perhaps, the architect planned only a two-story building and figured the strength of its component members according to the stresses that so light a weight would put upon them. Then perhaps the owner found the two-story structure too small for his purposes and ordered the construction of a third story. This new floor must be complete and identical with the floors below; but the architect must provide for a strengthening of the foundation and the bracings to accommodate the added weight. So the building might rise, story by story, and each addition would require the strengthening of the substructure.

A truer analogy, perhaps, would be an expanding manufacturing enterprise. The company starts with a factory in the suburbs and an executive office in the financial district of New York. Eventually it branches out and establishes a new factory in another section of the country. This factory is a complete unit so far as its technical processes are concerned, but the original executive office in New York, by increasing its staff somewhat, can direct the business of both plants and keep them supplied with orders and raw materials. And as the business grows the company keeps on adding new factories to its string, while, to conduct the business for all its plants, the central managing organization expands in corresponding ratio.

The A. E. F. grew by army corps, each corps a technical whole for actual combat against the enemy. But as corps after corps crossed the ocean and added its weight to the American strength at the front, the central management, consisting of a general headquarters and the whole multiplex array of activities embraced in the supply and support system of the expedition, had to expand in like degree.

The problem of planning out on paper a great expeditionary force was complex beyond the power of any single mind to grasp. The construction of the A. E. F. was in the hands of its own staff organization, the most competent intellects which the Army had at its disposal. Some sort of working plan was necessary. Our capacity in the United States to train and transport combat troops might be unlimited; in fact, we might have entered the war with millions of men trained and ready to go into the trenches to meet the enemy. But it would have been utterly useless to train these troops and send them across the ocean, unless we could accompany them with adequate numbers of supply troops. The builders knew roughly that for each corps there should be 135,000 combat troops, 15,000 general troops for the Army and corps commands and the line of communication, and 50,000 Services-of-Supply troops. They might specify in advance the combat troops with fair accuracy, but they could scarcely foresee all the kinds of units

From An Official Motion Picture

HUNDREDS OF SHIPS CROWDED LIKE THIS, AND EVERY MAN IDENTIFIED

Photo by Signal Corps

THE *MADAWASKA* TAKES A CROWD, JUNE 30, 1918

Photo by Signal Corps

LOOKING FORWARD ON U. S. A. TRANSPORT *MERCURY*, JUNE 30, 1918

Photo by Signal Corps

LOOKING AFT ON *MERCURY*, JUNE 30, 1918

needed in the Services of Supply; for the maintenance of a great expedition called for thousands of widely various activities, scattered through almost the entire range of human enterprise.

In order to build the A. E. F. scientifically, to provide the proper balance of troops sent to France and to avoid the fatal blunder of shipping any one class of soldiers in numbers out of just proportion to the whole; in order, further, to forestall confusion and give the simplicity of system that comes from dealing in small numbers, General Pershing and his assistants marked off the construction of the overseas army into definite periods of increment. For want of a better name, each increment period was called a phase. The word "period" is not an exact term to apply to the system, for it connotes time, whereas the working plan for the construction of the A. E. F. was but vaguely related to the time required. General Pershing's Shipping Schedule No. 1 made provision for six phases of this construction. Each phase, when completed, would add to the A. E. F. one entire army corps plus its necessary maintenance troops.

On October 7, 1917, this great plan was set down on paper and forwarded to the United States. The six phases provided for an overseas American Army of thirty divisions (five divisions to the corps). Counting in supply troops, the force thus projected would number 1,200,000 men. No time was fixed within which America was to complete the schedule. General Pershing hoped we could accomplish the feat by July 1, 1919—in twenty months. We can measure the increase in our ability to train and transport troops by the fact that in August, 1918, less than eleven months after the date of the Shipping Schedule, the Embarkation Service was forwarding to France the final units of the sixth phase.

After that the Army abandoned the numerical phase and substituted the monthly phase: we had become able to transport to France in a single month an entire army corps with its supports! The sixth was the last of the numbered phases. September, 1918, was an interval during which we shipped miscellaneous phase troops pushed out of their sailing priorities by emergency calls from the front; and then we began on the new system, naming each phase after the month during which it was being shipped. When the armistice came in November, we were still shipping troops of the October phase, the influenza epidemic and the assurance on November 1, 1918, that the war was about to end having slowed down the movement during the last eleven days.

One is not to suppose that the Army made a clean job of shipping phases of the A. E. F. to France. Frequently, before one phase had been completed there was a need overseas for troops of the succeeding phase; and it was not unusual for units of even three phases to be crossing the ocean simultaneously. Phases always overlapped somewhat. The Shipping Schedule, after all, was but a working outline, to be followed only so far as events warranted.

In the compilation of Shipping Schedule No. 1, General Pershing adopted the item number as the means of identifying troops arriving in France in response to the requisition. In the paper army thus laid down—some of it months before its human members had even been inducted into the military service of the United States—the prescribed units were assigned item numbers and initials.

The initial I stood for Infantry, E for Engineers, M for Medical Corps troops, Q for Quartermaster troops, and so on. The numbers, too, had meanings. Divisional troops bore item numbers ranging from 1 to 100 serially within each phase; troops attached to army corps bore numbers from 101 to 200; troops acting in the service of the commands of field armies, from 201 to 300; units for service in that great expeditionary institution known as G. H. Q., 301 to 400; and troops for the lines of communication, 401 to 500.

We can now read meaning into the item numbers assigned to the 34th Regiment of Engineers. Back in the early autumn of 1917, the A. E. F. builders had reached the plan for the third phase of increment. There they foresaw that the A. E. F. would need, at a certain point in its structure, a supply train and machine shop. In Paragraph No. 8 under Phase 3 they set down these requirements and gave them the item numbers E-403 and E-404. The initial E showed that the troops would be Engineers. Since the numerals were in the 400's, they indicated to anyone familiar with the system that the units were to serve within the lines of communication.

Months later the Operations Division of the General Staff, engaged in turning the abstract schedule into a specific force, came to Phase 3, Paragraph 8. The General Staff officers in Washington surveyed the troop resources which had now sprung into existence at the training camps; and at Fort Benjamin Harrison they found the 34th Engineers, three of whose battalions operated a supply train and a machine shop. These battalions fitted the requisition. Therefore, opposite the two item numbers of Paragraph 8 the Operations Division set down the name of the 34th Engineers. This act gave the regiment its place in the priorities for sailing. Thereafter the item numbers became an integral part of the regiment's name, to appear in every order concerning its travel. At length the 34th reached France. No need there for the A. E. F. debarkation officers to inquire into the regiment's qualifications and then find a place where it might serve: that place had been fixed months before. The debarkation officials needed only to turn to Shipping Schedule No. 1 to learn precisely why the A. E. F. needed the unit and where it was to be sent.

This system of identification was followed rigidly in the upbuilding of the first six army corps of the expedition, in so far as the needs of the force could be foreseen a year in advance. All combat organizations, at any rate, and many supply units which moved to France during the major part of the embarkation period, possessed item numbers assigned by the command of the overseas force in its first schedule. But at that early date General Pershing's experts could scarcely anticipate the enormous expansion of the need for units of kinds previously unknown in our military service. Finite minds could not be expected to lay down in advance every detail of a modern army as large as the A. E. F.

Who could know in the summer of 1917 that, less than a year later, our expedition would be calling for whole companies of meteorological experts to form a weather bureau for the A. E. F.? In many ways the accurate prediction of weather conditions was of invaluable aid to the Army. It enabled the Air Service to lay its plans with intelligence. Weather forecasts controlled to a great extent the employment of poisonous gases, for certain of these could be used with most success only when the wind blew toward the enemy. The draft reached out and took men from every calling. The training and experience of every one of the 4,000,000 men in the American Army was ascertained and catalogued in the great index of human talents compiled in the War Department in Washington. For the Operations Division to supply a whole company of weather prognosticators on short notice became a mere trifle. As the A. E. F. expanded, there were dozens of other calls for special units strange to our former military practice. Requisitions came for companies of typesetters and linotype operators, embalmers and grave registrars, motion-picture camera operators, cold storage experts, pharmacists, and coffee roasters and tasters—to name only a few of the innovations.

Nor could it be foreseen in the summer of 1917 that the Army would add whole services to the group it had known before—such organizations as the Chemical Warfare Service and the Tank Corps. In short, the phase program as laid down by General Pershing in October, 1917, did not begin to meet the developing actualities. Then, too, the putative army as it existed on paper in the Shipping Schedule was by no means complete, even in well-recognized troops. Often it became necessary for the A. E. F. to go outside its schedule in the call for such soldiers as quartermaster troops and motor transport troops. Many organizations proceeded to the seaboard and thence to France duly identified with item numbers from the original schedules; but thousands of others crossed in response to supplementary requisitions, and these units bore no item numbers.

The Operations Division tried to give each unnumbered unit identification by writing into all orders which concerned it, in juxtaposition to its unit name, the serial number of the Pershing cablegram which requested it and the cablegram paragraph number. But this plan did not work. The debarkation officers at French and English ports might or might not have before them copies of General Pershing's cablegrams. Seldom did the commander of an overseas unit know for what purpose he and his men had been summoned to France. For a time there was great confusion in France because of incomplete identification of arrivals; hundreds of individuals and organizations became lost, and some never did reach the posts for which they had been requisitioned.

A new supply warehouse might be set up somewhere along the lines of communication in France. The Q. M. organization discovered that the phase plan had not contemplated this establishment and had made no provision for troops to man it. The commander of the depot thereupon made requisition to the headquarters of the A. E. F. for, let us say, 200 quartermaster troops. This requisition in due time found a place in one of General Pershing's cablegrams. Perhaps at the time the Operations Division could find no quartermaster company numbering 200 men, but discovered one of 125 men. Naturally it dispatched the 125 men to France and made up the deficit in later shipments. Suppose the first organization, the one with 125 men, were lucky enough to meet at the French debarkation port a copy of its requisitioning cablegram. The debarkation officer sent the unit at once to its correct destination. But the succeeding detachments were almost certain not to be identified upon arrival overseas, and the port officer there would send them to any spot which needed such troops. Presently the depot commander would complain bitterly that his requisition was not honored. Operations, in Washington, could only reply that they had sent him his men long ago.

The plight of individuals whose services were not contemplated in the Shipping Schedule of October, 1917, might be even more unfortunate. A casual officer specially requested by the A. E. F. did not often know for what purpose he was being summoned, or where he was to go in France. No more would the debarkation port know about him; and he was usually sent wherever the A. E. F. port official fancied he could be of service. At one time a highly trained technical man, commissioned in the Army, was requested for special service in France. The A. E. F. officer who sent for him had in mind an important post which he was eminently qualified to fill. This casual lost his identity in transit. The service that requisitioned him searched France for him in vain for months—and found him after the armistice bossing a gang of stevedores at one of the army docks!

On April 24, 1918, the Army took measures to end this confusion. After that date the A. E. F. assigned item numbers to all excess organizations and officers summoned to France, and these numbers were outside of and different from the phase numbers of the original schedule. The lid was screwed down on embarkations except upon cabled requisitions giving item numbers. The A. E. F. adopted new arbitrary number groups for these excess troops. This system continued throughout the heaviest period of embarkation. It saved the A. E. F. a confusion that would have become tremendous after the late spring of 1918, when troops began avalanching on France.

By April, 1918, the number of recognized classes of troops had greatly expanded. The A. E. F. then adopted a system of initials to go with item numbers and help identify organizations. The list of abbreviations, as it stood at the time of the armistice, was as follows:

X	Reinforcements
K	New Units
W	Exceptional Replacements
R	Automatic Replacements
I	Infantry
CA	Coast Artillery
FA	Field Artillery
C	Cavalry
E	Engineers
A	Aviation

T	Tank Service
M	Medical Corps
S	Signal Corps
O	Ordnance Corps
MAR	Marine Corps
Q	Quartermaster
TRANS	Transportation Corps
G	Chemical Warfare Service
L	Light Railroad and Roads
F	Construction and Forestry
MOTOR	Motor Transport Corps
AG	Adjutant General's Dept.
IG	Inspector General
JA	Judge Advocate
SERVICE	Army Service Corps

An interesting development of item numbers occurred in the system of identification for replacement troops. Under the army maintenance plan adopted in the summer of 1918, replacements became automatic monthly. The authorities on this side of the water, in other words, did not have to wait for cabled requisitions for replacement troops, but shipped them at a predetermined monthly rate. Yet it was essential to prevent the identity of replacements from being lost and the men themselves from being diverted by the French debarkation ports to any other purpose than that for which they were shipped. Consequently the replacements were given item numbers, which the Operations Division assigned as soon as replacements reached camp and were organized into units. Other troop units did not receive item numbers until there was a special call for them from France, or until the Operations Division reached the places in the Shipping Schedule where they fitted. A replacement unit received its item number immediately after organization, and this item number ended in the initial R, which, as the above table shows, indicated that they were automatic replacements. A unit of soldiers arriving in France under the item number I-1119-R would at once be identified there as a company of automatic replacement infantry. The A. E. F. possessed the list of replacement numbers representing troops that were to reach France each month, and sent to the ports of debarkation instructions as to the destination of each body of replacements.

Just as, after April, 1918, all excess organizations received item numbers, so did all casual officers proceeding overseas for special work in the A. E. F. The casual officers' item numbers began with the numeral 1001 and proceeded upwards serially.

CHAPTER XIX

THE PROCESS OF EMBARKATION

NEAR the end of each month there was a meeting in Washington to determine the number and identity of troops to go overseas during the ensuing month, and to fix the order of their sailing.

The Operations Division of the General Staff called and conducted the meeting. The Chief of Embarkation brought an estimate of his port facilities for the accommodation and handling of troops during the period under consideration. The embarkation director stationed in America by the British Ministry of Shipping came with schedules of the names and passenger capacities of the English transports which were to sail from New York and Canadian ports. Our own navy representative was more cautious, but he undertook to predict what American transports would depart during the first ten days of the month, promising supplementary and additional schedules each succeeding ten days. The Operations Division had before it the A. E. F. Shipping Schedule, the names of organizations assigned to the schedule, and data covering the readiness of these organizations for foreign service. The representative of the troop-movement office was present to see to it that the arrangement of the monthly sailing priorities was coördinated with the efficient employment of the available railway equipment.

Anyone who can add two and two together can see that, with all this information and ability encompassed by the walls of a single room, it was entirely possible to project on paper a complete diagram of troop embarkation for a month in advance. We knew to a man how many passengers the ships could carry and to a man where the passengers were coming from. Therefore, to any lover of the orderly, the systematic, the thirty-day sailing priority schedule that eventuated from this meeting was a thing beautiful to see; it was System with a capital S.

Unfortunately for all planning, human ability is fallible and inanimate objects are perverse. Transports had a way of sinking with gaping wounds in their sides where German torpedoes had struck in. Storms and head winds upset the best prognostications of the mariners. Ship machinery broke down and had to be repaired. Labor troubles delayed vessels in the foreign ports. Bunker coal was not always available on the minute. Never once did the actual sailings of British transports befall as the predictions had said; never once did the American troop carriers depart seaward in exact accordance with a ten-day forecast. Events showed that the basis for the monthly sailing schedule was more or less accurate guesswork.

It was evident that the embarkation machine must have a governor like an engine's; a contrivance for opening the valve wider to admit more troops when the sailings exceeded the estimates, and for closing the throttle when the influx of soldiers threatened to flood the port. The embarkation camps, a reservoir that could accommodate, in a pinch, 60,000 troops, aided in this adjustment; but they alone were not sufficient to make the sailing schedule work. The adjusting mechanism, the governor on the machine, was the dispatch office at Hoboken. It fitted the prospectus to the actual day-to-day conditions. A small handful of officers did the work of the dispatch office, and they were busy men—how busy an illustration will show.

The day was a Thursday. The business in hand at the port was the embarkation of 32,000 troops who were to sail on Saturday on a British convoy of fourteen vessels. The 32,000 were all in the embarkation camps (along with other troops awaiting sailings, the total filling the camp to capacity), undergoing the final preparations for the voyage. The dispatch office regarded the job as done; it was now working in terms of the future, and had already ordered to New York 32,000 troops to occupy the barracks to be vacated on Saturday by those who were embarking on the British ships. At this juncture the Brit-

Photo by Signal Corps

AT ALPINE, WAITING FOR FERRYBOATS

Photo by Signal Corps

BOARDING FERRY FOR PIERS

Photo by Signal Corps

LANDING FROM FERRY AT HOBOKEN

Photo by Signal Corps

ENTERING PIER FROM RIVER END

ish embarkation officer telephoned that the British Ministry had cut four ships out of the convoy, reducing its capacity to 25,000 troops.

Such belated changes were not unusual. Frequently the home conditions in England demanded the temporary withdrawal of ships which had been assigned to the American Embarkation Service. Once in the spring of 1918, after the British Food Controller had inventoried the reserves and found only six days' supply of food in all England, the Ministry drastically cut down the passenger space on the transatlantic ships and crammed staterooms and berth decks, as well as cargo holds, with food for the English people. It was some such emergency which had arisen now. The dispatch officers took the upset with philosophical *sang froid*. They were used to it. They could make the adjustment by sending out telegrams to catch 7,000 troops at interior camps before they had entrained for the rail journey to New York, where now there would be just so many barrack accommodations fewer than had been expected.

On Friday the British withdrew another ship, this time a big one. On Saturday they cut out another, and the mutilated convoy to sail that day contained only eight ships instead of fourteen, with berths for 16,000 soldiers instead of 32,000. It was a jolly situation. Actually on the piers was an excess of men, brought there during the night, before the final ship was canceled; and traveling steadily toward New York were troop trains on which rode nearly 16,000 men more than could find places to eat and sleep in the embarkation camps.

Such *contretemps* as this kept the dispatch officers from participating in that popular Hoboken pastime, wondering when an ingrate Government was going to send one to France to see some of the excitement. After a man had finished a twelve- or fourteen-hour stint during which nothing ever moved smoothly or as expected, he was too tired to wonder about anything except how long it would take him to get to bed. The dispatch office was the contact point between theory and conditions, and the job of absorbing the friction was no picnic.

Yet the managerial acrobats of the office were never found wanting. They solved every problem, and they met the situation set forth in the preceding paragraphs. In that instance they gave orders to the port railway agent to slow down the incoming troop trains to the speed of freight trains, so as to utilize the coaches as mobile barracks; and they postponed further travel toward the port until the almost daily sailings had cleared away the congestion.

The dispatch office, when it began its career, was part of the Port Commander's own office. In those days a troop convoy sailed about every two weeks. The dispatch officers then both ordered troops to come to New York and conducted them through the port. As the volume of travel grew, these duties became too heavy for a single agency to handle and were divided among other port organizations. The whole port system expanded and ramified in hundreds of directions; it came to embrace twenty-one distinct departments, each administered by a branch of the port organization. The dispatch office was then not even a branch: it was the twig of a branch. But, though humble in position, it was mighty in importance.

The dispatch office was the Port Commander's right hand in his contact with all traveling organizations. When a unit was first placed on the sailing priority list, the dispatch office notified it of the fact and sent to it the booklet *Embarkation Regulations* and other general instructions relating to its travel. When, later, the unit commander telegraphed to Hoboken for travel instructions, the dispatch office, in the name of the Commanding General, replied, designating the embarkation camp, the time of arrival, the equipment to be brought and that to be left behind, and other steps to be taken. When the unit was in the embarkation camp, inspected, equipped, and ready for embarkation, the dispatch office ordered its advance party to the transport and named the day for the unit itself to move to the designated pier.

In general, Washington expected the Port to follow the fixed sailing priorities; but the War Department was not hidebound by its schedules. The administration, at least informally, permitted a certain latitude to the dispatch office; it was content to accept judgment of the men who were on the spot and confronting the actual conditions. If the Port were reaching out desperately for troops—as it was at the end of the war, when the transport facilities were exceeding the ability of the country to raise and prepare troops for foreign service—the dispatch office skipped about in the priority list and took any troops who were ready for departure, rather than hold up the whole column because one organization or another had encountered delay in getting away from its training camp. It is to the credit of the organization in Washington that it invariably backed up and authenticated such deviations from the orders.

The flexibility of the Embarkation Service was never better demonstrated than in the spring of 1918, when the British first placed their transport tonnage at our disposal. Up to that time we had been sending over divisions as nearly complete as possible—infantry, machine gunners, artillery, and divisional support troops—always keeping two, three, or even four divisions proceeding overseas simultaneously. Also as nearly as possible, we were completing the phase shipments in regular order, so as to place in France an entire army corps, with its necessary communication troops, before proceeding to ship any of the next corps. Exigencies at the front upset this plan for a time. The Germans were driving ahead and consuming the Allied trench troops at a ruinous rate. The British offer to transport, feed, and brigade with the B. E. F. six divisions of American line troops, and when these had been supplied to take another six, was an offer to carry infantry and machine gunners only. The British had plenty of artillery to support a retreating action. Later on these brigaded divisions were to join the A. E. F., and only then were they to receive their artillery.

The change in embarkation plans came suddenly in March, 1918. Washington telephoned that until further notice the Port was to ship only infantry and machine gun troops to France, whether in British or American convoys, and that only if the supply of available infantry organizations and machine gun companies ran out was the Port to send artillery and Services-of-Supply troops. At that time Hoboken was engaged in the routine embarkation of the Third Division. This shipment it abruptly broke off; and for weeks thereafter ship upon ship departed from New York loaded to the rails with doughboys only. Later on, when the weight of these troops began to be felt by the enemy and the emergency was over, the Port resumed its work on the Third Division.

The dispatch office almost always succeeded in keeping the embarkation camps crammed to the limit; and for whole months during 1918 there was not a day when a man with telescopic bird's-eye vision, looking down on the railroad system of the United States, could not have seen, in motion toward the seaboard, trains carrying from 15,000 to 20,000 more troops than the port camps could hold if anything happened to curtail embarkation. In the face of this nice adjustment of troop-flow to sailing space, last-minute changes in transport or convoy capacity were chronic. Sometimes, to avoid holding a ship in port for repairs, it was decided to make repairs during the voyage. Then bunks had to be torn out to make space in which to stow the repair parts, and the passenger room was reduced by so much. The British seldom adjusted passenger and cargo space on a ship long before the eve of sailing.

A last-minute change in passenger capacity might be slight, but for the dispatch office it often created a problem quite out of proportion to the change. Military law required every organization to cross the ocean undivided. A change that threw one company of a regiment off a transport, automatically threw off the whole regiment, and the dispatch office had to hunt up, at the eleventh hour, a new unit to fit the reduced space. Casuals and supply troops came in handy at such times: they usually traveled in small units, and consequently packed well.

In spite of all the difficulties in providing troops, only one transport sailed from America with any empty berths, and even that one was three-quarters filled. It is true that, during

From An Official Motion Picture

COFFEE AND ROLLS AT RED CROSS PIER CANTEEN

Photo by Signal Corps

FIRST FOOD SINCE 3.00 A.M.

From An Official Motion Picture

"SAFE-ARRIVAL" CARDS SLIPPED INTO CAPS BY WELFARE WORKERS

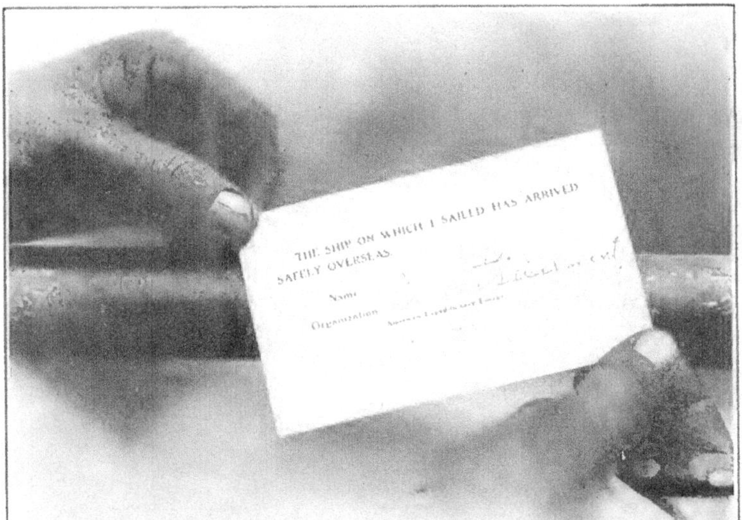

Photo by Signal Corps

A "SAFE-ARRIVAL" CARD

the influenza epidemic, ships sailed with fewer passengers than they could carry, but this reduction in capacity was insisted on by the medical authorities. The dispatch office regarded a ship as loaded full when its troops totaled the capacity allowed by the army doctors.

The Port not only loaded ships to capacity: it overloaded some of them. Several of the former German vessels, originally built to transport troops of the German army, carried loads far beyond the utmost capacity which their builders had reckoned on. We installed berths far more thickly than any foreign plans had contemplated; and, furthermore, we placed soldiers aboard in such numbers that there were not enough berths for all, and the men had to sleep in shifts. Never in history had such an expedient been resorted to.

The dispatch office controlled overseas troop movements to New York's subsidiary ports, from each of which, Boston, Philadelphia, and Baltimore, two or three ships sailed in an average month. Occasionally, too, there were shipments from Portland, Maine, and from Montreal and Quebec. At these ports there were no embarkation camps; the troops were equipped and inspected for overseas service at their home camps before proceeding to shipside.

All troops came to the New York piers by ferry. The water trip from Camp Merritt to the North River has been described. Troops at Camp Mills or Camp Upton, on Long Island, traveled by train to the waterside terminal in Long Island City and there embarked on ferry-boats. As the men scrambled to the floor of the pier they lined up to respond to the company roll-call. This formality ended, they were allowed to approach the tables set by canteen workers of the Red Cross. Here, in cool weather, rolls and coffee were served to them; during the summer, ice cream and cold milk.

The Red Cross Canteen Service at the Port of Embarkation was at first an organization of four women, Mrs. Roy Rainey and Mrs. J. Ellsworth of New York and Mrs. Palmer Campbell and Mrs. Franklin Hart, Jr., of Hoboken, New Jersey. At the time of the armistice the service mustered over three hundred devoted women, of whom a hundred and twenty were stationed at the piers of New York, sixty at Newport News, and the rest at Boston, Philadelphia, Baltimore, Montreal, and Halifax. The women who enlisted for this service, many of them New York society women unused to rough work, agreed to give their full time. The life was arduous. It required all canteen workers to be on the piers and ready for business at five o'clock in the morning, which meant rising at four. It is of record that throughout the entire embarkation not one of the women ever reported late.

The soldiers on the piers consumed enormous quantities of supplies. The Red Cross installed at Hoboken the largest coffee-distilling plant in the world, and then put in a bigger one at the Chelsea piers on the New York side of the river. A single day's embarkation at New York often called for the serving of ten tons of hot liquid coffee, three tons of rolls, about a ton of cigarettes, and several hundredweight of ice cream and cookies. This refreshment was most welcome to men who had spent the night on their feet. Many a time it turned a cold gray dawn into a rosy one. The Port of Embarkation paid official tribute to this service for its wholesome effect upon the morale of the embarking soldiers.

After the men had visited the refreshment tables, the Red Cross and Y. M. C. A. workers went among them with the "safe-arrival" cards. These were uniform, printed post cards for the soldiers merely to sign. Each read: "The ship on which I sailed has arrived safely overseas." No date appeared, nor was the signer permitted to write in the name of his organization. The soldier signed and addressed as many of these cards as he chose, and kept them on his person to deposit in the military mail bag which he would later pass at the head of the gangplank. Just before the ship sailed, the mail bags were taken off. The Hoboken military post office bundled the "safe-arrival" cards together and held them pending the receipt, a week or so later, of the cablegram announcing that the ship had reached Europe. The cards were then forwarded through the regular mail.

In the ship mail bags the soldier might deposit, not only these cards, but also whatever final letters he might wish to write. If he desired his letter to go through to its destination immediately, he deposited it unsealed. It then went to the office of the port censor, who read it and, if it contained no contraband information, forwarded it at once. Sealed letters the censor did not read, but he did not forward them until the transport had reached England or France. Commissioned officers and army nurses were permitted to write out "collect" telegrams announcing safe arrival overseas. These they left at the port post office to be filed at the telegraph office when the transport reached the other side of the ocean.

When convoys began sailing almost daily from New York, one result was a huge accumulation of "hold" mail—mail which must await advices from Europe before it could be forwarded. The military post office devised a system for handling it so as both to preserve the secrecy that then shrouded ocean travel and to maintain the identity of mail and prevent its loss. Just before a ship left a pier, her mail bags were collected. The unsealed letters were then sorted out for censoring and forwarding. The rest was sealed in sacks, and all the sacks were marked with the transport's number. We have seen that each troop transport went by number rather than by name. *Leviathan* was No. 22. Mail bags taken from the *Leviathan* were marked No. 22. When the cablegram announced that transport No. 22 had again passed safely through the war zone, all the No. 22 mail bags were sent to the regular post office.

At Hoboken there was only one instance of a soldier attempting to send out information calculated to be of aid to the enemy, and even in this case it is questionable if the man's actions were criminal or only foolish. There was a positive and well-understood order at the piers that all letters must be dropped in the gangplank mail bags and nowhere else. This man wrote a letter and paid a pier stevedore a fee to post it in a city mail box. The Intelligence Service at Hoboken discovered his action before the letter reached the addressee. The

New York post office turned its forces loose on a hunt which presently found the missive. It proved to be a letter addressed to the soldier's wife in Brooklyn, a woman born, as investigation showed, in Germany. The letter contained a full list of the troops on the transport and even named the English port to which the ship was bound—probably a lucky guess, for all convoys sailed under sealed orders, and even the convoy commander did not know his destination until well out at sea. The soldier was arrested and taken from the ship. He protested his innocence, declaring that he was unaware of the regulation requiring all mail to be deposited in the ship mail sacks. It was impossible to prove a case of espionage against him; but he was tried by court-martial for violating the order pertaining to the mailing of letters, convicted, and sentenced to serve a term in the penitentiary.

At the piers the medical officers made their final inspections; and sometimes, even at this last hour, they took out men suspected of disease. Personnel officers took the opportunity to make a final inspection of the nationality of embarking soldiers; and now and then they found an alien, or even an enemy alien, who had slipped through all previous inspections.

Presently the force of gangplank checkers came on duty. Meanwhile the pier laborers had brought to the feet of all gangplanks tall desks at which men could work standing up. Each company marched to the gangplank by which it was to board the boat. The company commander took his place at the desk, the men's service records in a case before him. In front of the desk stood the port checking officer; at his elbow, the first sergeant of the company. Embarkation proceeded by squads. The command was right or left by files; and, a squad at a time, the men approached the gangplank. Passenger list and service records were arranged to tally with the arrangement of the men in their squads. When a squad halted before the desk, the checking officer called out the first name on the passenger list. The first man of the squad, when his name was called, responded by repeating his name, family name first,

From An Official Motion Picture

CHECKED AGAINST COMPANY RECORDS

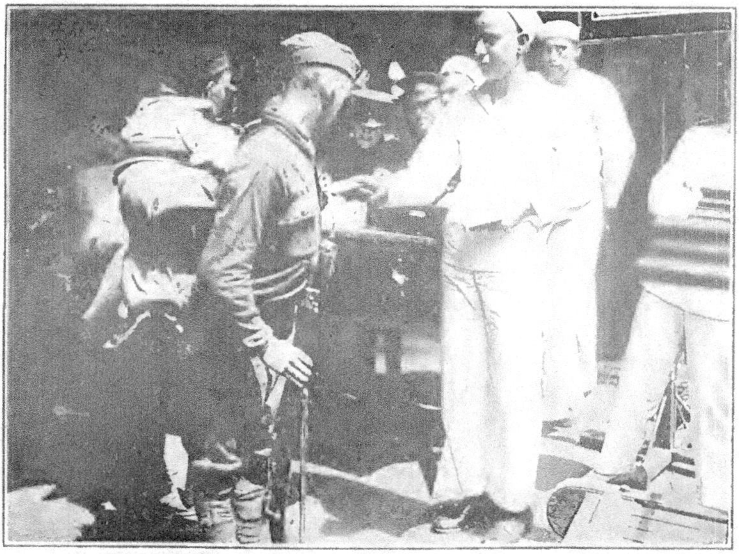

Photo by Signal Corps

SHIP BILLET CARDS AT FOOT OF GANGPLANK

Photo by Signal Corps

FRESH ARRIVALS FROM EMBARKATION CAMP

Photo by Signal Corps

BOARDING SHIP

given name and middle initial afterwards. He had previously received instructions to speak loudly and distinctly. He was then ordered to ascend the gangplank, and the checkers called the next name. If any soldier were absent at embarkation time or were withdrawn for any reason, his service record was taken from the company file and retained by the embarkation officers.

As each man passed the desk he received a billet card which noted the compartment of the ship in which he was to be quartered and the number of his berth in that compartment. At the top of the gangplank he met a member of his organization's advance party, who escorted him to his bunk. There he was told to remain until the order came which permitted him to go on deck. During the greater part of the embarkation period, soldiers were forbidden to appear on deck until the ship had reached the rendezvous in the lower bay. Officers were ordered to remain with their men during this interval. The sleepy men usually beguiled the enforced imprisonment by turning in for much-needed rest.

All through the early months our transports went down New York Bay without a soldier visible to those aboard ferries or other harbor craft or in the windows of the tall office buildings of lower Manhattan. In the late summer of 1918, the Government gained complete confidence in its ability to send transports safely past the enemy submarines. Moreover, to depress the enemy morale, the United States actually wanted the German Government to know the high rate at which we were sending troops to France, and for that reason it made no attempt to conceal the overseas movement. Each transport went down the river crowned with olive drab. Cheering soldiers, thick as swarming bees, encrusted her rails, lifeboats, ventilators, even her rigging; and the regimental bands played quicksteps on the decks.

The final act in embarking a company was to gather up its baggage detail and send the members aboard. Each unit, on its arrival at the pier, sent a few men to make sure that all company baggage was loaded. This work frequently continued up to the minute of sailing. Meanwhile the company commander remained at the gangplank desk to check aboard the baggage detail. After these last individuals had passed aboard, the port checking officer took all service records to which there had been no men to correspond, so that if the stragglers reported later they would find records waiting for them. The company commander then certified in writing that he possessed records for every man embarked. Even on ship the commander kept the record cards handy, lest at the last minute it might become necessary for the Port to take off some member of the command. During the influenza epidemic, hundreds of men suspected of having the sickness were removed, even after the ships had gone down to the lower bay.

Casual officers were checked on the transports in much the same way, except that the port organization itself made out passenger lists for them and prepared their credentials.

The pier organization grew extremely adept in its work. It could easily load 1,000 men in an hour. The port record was 250 men checked aboard at one gangplank in seven minutes. The greatest day in the history of the Port was the twenty-four-hour period beginning on the morning of August 31, 1918. In that time the port organization loaded more than 51,000 troops on seventeen transports. This was the largest number of passengers, either civilian or military, that ever sailed from any one seaport in any one day in the history of the world. An appendix* to this narrative shows the transport sailings from New York from the 1st of July, 1918, to the date of the armistice. It will be noted that sixteen vessels sailed from New York on August 31 and September 1; and the *Aquitania*, listed as sailing September 2, but actually loaded on September 1, really belongs within the record twenty-four-hour period.

These record embarkation days, however, were not to be the real test of the port organization's mettle. Its ordeal by fire was the interval between the 1st day of November, 1918, and the hour when the armistice went into effect. Scarcely any troops at all were embarked in those ten days; yet they were

* Appendix E.

the most memorable in the history of the Port and the time of its most signal achievement.

On the morning of November 1, a select few of the highest army and navy officers in Washington were admitted to a most startling military secret. Our Intelligence Service abroad had gained two pieces of information of supreme importance. The first was that the dominant party in Germany was ready to sign an armistice, no matter what its terms. But the matter was not settled yet, for the bitter-end element favored a last desperate measure that might yet turn back the tide of defeat. This measure was nothing less than to throw the German navy at the Grand Fleet of the Allies and America. Both parties in Germany had agreed upon this course. In fact—and this was the second piece of information—the German Admiralty had sent the order to its fleet to go out to victory or destruction.

The Secretary of War at once decided to stop further embarkations of American troops for France, if that could be done without a word of the truth leaking out. There were several important reasons for such a course. Of these, the financial was the slightest. The Government was not pinching pennies, and the cost of transporting another 100,000 or so men to France made little difference, although it was an element in the situation. A graver matter was the impending demobilization. As peace loomed, Washington realized that it had a problem on its hands—to bring back the A. E. F. More than half that great force had crossed the ocean on British and other Allied ships. Upon the conclusion of an armistice, the British Empire would immediately withdraw her ships from our use and put them to work returning to their native shores her own home and colonial armies. We should be left to bring back our own men in our own ships; ships which, loaded to capacity, had taken more than a year to transport to France less than half of the A. E. F. Therefore the addition of even a single man to that force would make our problem of demobilization so much the harder.

But both these reasons together would not have warranted the danger incurred by the Government's decision to stop embarkation. There was another overmastering consideration—a paramount one which, taken alone, would have justified the step. In the light of history, we have to-day the comfortable knowledge that the battle order to the German fleet was met by a mutiny which touched off the German revolution and cast out the Hohenzollerns. But it must be remembered that on November 1, 1918, every one of the army and navy heads who had received the secret tidings expected that the greatest naval engagement of all time was at hand, if it had not already begun. The Government had every confidence in the ability of the Grand Fleet to meet the issue; but there was unquestionably the possibility that some of the German ships of war might win through the Allied naval cordon and reach the open Atlantic. In that event every American transport at sea would be in terrible danger. We did not protect our convoys at sea against armed battleships. The escort was protection against only a chance merchant raider of the enemy and against submarines. An enemy cruiser could have worked havoc in one of our troop convoys. When you consider that our embarkations, even at the average rate, meant as many as 150,000 American soldiers on the ocean at once, you can imagine the anxiety of those in Washington who knew that a naval battle was imminent. To stop embarkation would keep tens of thousands of American boys out of this danger.

Yet, if it were to be stopped, the action must be taken in complete secrecy—secrecy that would prevent the truth from becoming known, not only to the enemy, but to the Allies as well. The reason for keeping the information from the enemy is evident. The military party in Germany was still strong. If the junkers discovered that we had stopped embarking troops, they could argue that America was not so strong as she appeared; that she had seized the first rumor of peace as an excuse to bring to an end an effort which must evidently be exhausting her resources and strength. Such an argument might induce Germany to prolong the struggle.

No more did we care to let the British and the other Allies learn of the action. The escape of a German cruiser or two meant no such disaster to any one of the Allies as it might

Photo by Signal Corps

LAST LETTERS HOME BEFORE SAILING FOR FRANCE

From An Official Motion Picture

MAIL SACK AT HEAD OF GANGPLANK

Photo by Signal Corps

TROOP MAIL AT HOBOKEN HELD AWAITING ARRIVAL OF TRANSPORTS IN EUROPE

Photo by Signal Corps

A GANGPLANK LEADING INTO U. S. A. TRANSPORT *LEVIATHAN*

possibly mean to us. Washington, under the circumstances, regarded the stopping of embarkation as being our own business and a step justified by our trust in the reports from the Argonne that the enemy's line was broken and that the severing of his communications at Sedan was imminent.

On that momentous morning Brigadier General Hines, the Chief of Embarkation, was called into conference and asked if the step in contemplation were possible of achievement. He explained the difficulties in keeping secret so radical a departure. If only our own transports had been involved, the matter would have been simple; but more than half the troopships in our service were foreign, owned and operated by foreign companies—British companies for the most part. When these ships began leaving our ports empty, it would be most difficult to keep the operating companies from discovering the truth. Still, General Hines declared his willingness to accept the responsibility for the success or failure of the plan.

His instructions to the Port were so confidential that he did not dare trust them to the usual means of communication. He confided the situation to his chief executive officer, Colonel Waddell, and sent him to New York to communicate the news and the orders orally to the Commander of the Port, and to just so many other officers as had to be let into the secret. The orders were for the Port to proceed with every routine embarkation function as usual, stopping just short of checking the troops up the gangplanks. Ships should be sent out on schedule, but sent empty. German agents, if they were watching the port, would see nothing out of the ordinary.

There was but one outsider to whom the truth had to be told. This was Major P. A. Curry of the British Army, acting as New York embarkation officer for the British Ministry of Shipping. Our embarkation officers outlined the whole plan to Major Curry and obtained his coöperation, which included his promise not to communicate the truth to his home government. If London pressed him irresistibly for an explanation of why the ships were empty, then he had our permission to fasten the blame upon an alleged failure of the American Embarkation Service to deliver the passengers.

Of the agencies of the Port, the dispatch office was the only one admitted to the secret. The rest of the great organization rumbled smoothly on, serenely unaware that, for Hoboken, the war was over. Not even the entire personnel of the dispatch office knew what was going on. The two or three dispatch officers in the secret used to return to their desks at night after the clerical workers had gone, to write and send the telegrams which kept the port from being flooded with incoming troops.

At the time of the order to stop embarkations there were twenty-three troopships awaiting their loads in New York harbor. The advance parties of a number of overseas units had gone on board some of them. The port officers made one excuse after another to explain to the British captains why the troops did not come—this unit had been quarantined because of the influenza, that one was held up by lack of equipment. Sailing day arrived for a British convoy of seven ships; no postponement was thinkable. Major Curry went to the convoy commander and, with well-simulated disgust, told him to start out without passengers—the Embarkation Service had fallen down. The convoy sailed. It was then the discreet major's duty to cable his government the passenger lists, so that the English ports might make their arrangements for accommodating the troops. He was strangely derelict in performing this routine task. Day by day the Admiralty in London cabled more insistently for the lists. Major Curry remained uncommunicative until the convoy was but twenty-four hours off the English coast; then he cabled laconically that no troops had sailed on the convoy, and that the American War Department would explain. Not until after the armistice did the British Government learn that we had ceased to ship soldiers on November 1. The Distinguished Service Medal awarded to Major Curry by the American Government was largely a tribute to his discretion during the first eleven days of November.

On November 1 troops at two interior camps were about to

start for Montreal and Quebec, there to embark upon British ships. The Hoboken dispatch office canceled their travel orders by wire and notified the British ship captains not to wait for their passengers. It was safe to countermand embarkation orders to troops in the interior of the country, for it had often been done. The dispatch office was therefore able to stop all travel toward the port without giving anyone an inkling of the true situation. Fortunately, not many organizations were scheduled for embarkation at this time. The heavy movements of the summer had virtually exhausted the supply of trained troops in the United States. In late October the Port was embarking chiefly supply troops, labor battalions, and other organizations which did not require extensive training. At the end of October the Eighth Division, whose travel to the port from Camp Fremont, California, we have already cited; was making its way to France. The 8th Infantry, a member of the Eighth Division, was the last combat unit to go overseas. The other three infantry regiments of the Eighth Division were held in Camps Merritt and Mills and, after the armistice, sent to interior camps for demobilization.

It was more difficult to conceal the stoppage from troops resting in the embarkation camps. In the aviation camp at Garden City, Long Island, was a squadron of aviators most anxious to get to France. Three times they sent their baggage down to the port, and three times the Port sent it back again, each time with the excuse that the transport was held for repairs to her machinery. In spite of all precautions, the rumor went forth that something unusual was going on at Hoboken. When the order came to stop embarkations, the advance parties of several troop organizations were already on the transports. The port administration withdrew these parties and sent them back to the embarkation camps. The news of this occurrence ran swiftly to the New York newspaper offices. Within a short time a body of news correspondents besieged General Hines in Washington for an explanation. The general looked his questioners in the eyes and told them that they were on a false trail—that no troops had been pulled off any transports. Then the general privately called Hoboken by telephone and ordered the advance parties put back on the ships in a hurry, before the newspapers could check up the truth of his statement.

Fortune is said to favor the bold. The German sailors mutinied rather than face the Grand Fleet. General Pershing cut the enemy communications at Sedan. And the new German republic signed the armistice, thereby ending a suspense which was becoming unendurable in certain quarters in Washington. Any least slip-up, however,—a German victory at sea, a German stand in the field, even a partial enemy success that would have prolonged the fighting,—and those responsible for the cessation of embarkation, no matter how well justified their course, would have had to face the country with the admission that the American troop program had been set back two weeks or a month as a result of it.

CHAPTER XX

SOME NOTES OF TIDEWATER ACTIVITIES

SECRET panels that open to the touch of hidden springs, stairways concealed in unlooked-for places—we are in the realm of the mysterious. And not in the palace of an ancient intriguing Venetian doge, as you might suspect, nor yet in some mediæval Norman castle, but in the very heart of commercial America—in the offices of the Hamburg-American Steamship Company, at No. 45 Broadway in the city of New York.

When the Government, upon the declaration of war, seized the American properties of the German transatlantic lines, it fell heir not only to the system of modern piers on the New Jersey side of the North River, but also to the office building put up on lower Broadway by the Hamburg-American Line and occupied by it for a number of years. For several months thereafter the building remained tenantless. In November the New York branch of the War Trade Board moved into a section of offices on the street floor of the building; but the upper stories continued to gather dust and to echo only to the occasional tread of watchmen. Then the port organization, outgrowing its quarters at the Hoboken docks, sought space for expansion. It sent its construction forces to renovate the building and put it in order for occupancy. When the repair men came to examine the building, they made some discoveries.

With its cherry-red woodwork and walls tinted in strong shades, the building was somewhat more ornate than the usual American office building. It had spacious halls, high-ceiled rooms of generous dimensions, and broad windows looking out upon the bay and the mouth of the North River. The renovators found the chief executive office wainscoted to the ceiling in rich mahogany. As the workmen explored along the walls, their fingers touched unobtrusive buttons, whereupon panels swung open. One such door concealed a strong fireproof safe. Another opened into a hidden elevator shaft in which moved an automatic electric cage down to a basement passageway, which in turn led to a rear exit. By this route a man could leave office and building quite unobserved by possible watchers on the busy thoroughfare on which the building fronted. Still a third panel covered a dumb-waiter shaft connecting with a mezzanine kitchen above, which could be reached by a private stairway. An exploration revealed commodious pantries and refrigerators. It was evident that the former occupants of the executive office were prepared to stand a considerable siege.

In the spring of 1918 the Government occupied the whole of this building. The Navy took an entire floor. The important Shipping Control Committee occupied six floors. The United States Shipping Board, the United States Railroad Administration, and other official agencies all took space; and the Army Transport Service (later to be known as the Port and Zone Transportation Office) occupied the rest and hung its sign above the Broadway entrance.

The prime and obvious business of the Port of Embarkation was to embark troops and to ship supplies to France; but the successful conduct of this work entailed the prosecution of a host of related enterprises. There was the task of superintending the entire port personnel, a staff which at one time included close to 40,000 men and women. The inspection of troops and of their equipment created another major activity at the port. The Port had its own lawyers and courts. One set of advocates concerned themselves with purely military cases. Another set devoted their attention to the innumerable questions of admiralty law that arose in the operation of the transport fleets. The port surgeon administered camp and base hospitals, inspected troops for physical disability, received and transported the sick and wounded, and operated harbor hospital boats and hospital trains. He also assigned medical officers to the troopships. This complete travel agency, the Embarkation Service, even supplied spiritual and doctrinal comfort to its voyagers, by placing chaplains at the embarkation camps and on the transports. The port chaplain also supervised the many welfare activities at the port. Another section administered the storage of materials awaiting ocean transit. Another employed and directed the gangs of stevedores at the freight docks. A branch of the Port, with nearly 6,000 military and civilian employees, built and kept in repair the hundreds of structures which housed the organization. The great Port and Zone Transportation Office chartered the army transports, repaired the cargo transports and provided crews for them, secured and operated whole fleets of harbor boats, and rendered many other important services. Another organization made out the pay checks. There were boards of survey to estimate the deterioration of chartered ships, for the Government's guidance in its settlements with the owners. There were survey officers to fix the amount of damages done to ships and cargoes in accidents. In short, the New York Port of Embarkation was comparable in size to some of the largest industrial enterprises in the United States. If you include the crews of cargo transports and the forces of machinists and laborers at the various harbor repair shops and ship railways, the port activities gave employment to nearly 100,000 individuals.

It is beyond the scope of this narrative to describe the intricacies of the organization; to show how, as the work grew, various activities came together under certain heads, sprang apart again like the luminous particles in a kaleidoscope, and re-formed once more in other combinations. The port organization was no rigid institution. It kept experimenting with different forms of control in an attempt to approximate the ideal.

One of the earliest undertakings of the Port was to provide quarters for itself. The Port's construction division attended to this matter. Its first task was to build a group of structures in the yards at the Hoboken piers and to add a third story to a long section of the pier bulkhead building, all of this new space for occupancy by port headquarters and by branches whose work was at the piers. The Port Commander had his office in the bulkhead building. The constructors extended into the pier yards a railroad track which connected with the Hoboken Shore Railroad. Troops which came to the ships directly from their training camps were sometimes brought in to the piers on this track.

The construction division also had charge of the Army's harbor dredging. When the Government undertook to move the great steamer *Vaterland*, later renamed *Leviathan*, from her berth at Hoboken, it found that she was fast in the mud. The construction division put in dredges and in four days cleared out the slip. Thereafter the division operated a squadron of dredges that kept the army slips unobstructed. Eventually the port builders constructed a great development enterprise of erecting groups of piers and warehouses on the Jersey and Brooklyn water fronts and putting up barracks in various places in the metropolitan district. When the Army occupied several of the Bush freight piers in South Brooklyn, the port constructors fenced in that great terminal and flood-lighted the stockade so that skulkers by night could not approach undetected.

The Port and Zone Transportation Office managed the ferry-boats used in the transportation of troops within the jurisdiction of the Port, and also operated all other harbor craft used by the Army. After the armistice, the Port and Zone Transportation Office took over bodily a great part of the organization and duties of the Shipping Control Committee, including the function of loading military freight on transports.

The whole Atlantic commercial equipment of lighters, tugs, car floats, and small passenger steamers did not measure up to the needs of military shipping in 1918. The Embarkation Service bought and chartered all the suitable harbor craft it could find, and then found itself confronting the necessity of building more. There was no established industry to which the Service could turn and order new floating stock, nor was there any other branch of the Army which could supply it;

and the Service showed its versatility by going ahead and building its own harbor boats. In this work the Service recognized the world's desperate need of ocean tonnage. It proved its recognition by putting no additional demand upon shipyards which were launching hulls for the United States Shipping Board, for the Navy, or for any other government branch (and practically all existing American shipyards were engaged in such work). Rather, it allotted its contracts entirely to new yards created expressly to serve it.

Nor did the Service wish to compete extensively for steel, so vital to other war activities. Therefore it built just as few steel vessels as possible. It specified wood construction for the hulls of such small boats as junior mine planters, motor tugs, and troop launches; and then it stepped clear outside standard shipbuilding practice and adopted for the rest of its equipment the new reinforced concrete construction which was then beginning to intrigue the attention of the marine world. The Embarkation Service was not only one of the American pioneers in building boats of concrete, but it attained more success in this direction than any other large constructor, public or private.

Having complete military jurisdiction over all its harbors, the Embarkation Service was required to defend those harbors against possible enemy attack. The only attack anticipated was by submarines. Part of the harbor defenses were mine fields at the approaches; and the Service was presently operating a fleet of mine planters, large and small.

As the experience of the Embarkation Service broadened, it began to make greater use of port waters in the transportation of troops; and thus it removed from the terminal railways some of their war-imposed burden. Its success in bringing troops to New York piers from Camp Merritt by the Hudson River caused the Service to extend its use of natural waterways. Before the armistice came, experimental embarkation from Camp Lee in Virginia had used the James River as a highway between the camp and the piers at Newport News. One outgrowth of this policy was a scheme to build a fleet of river steamers designed for carrying large numbers of passengers.

The Embarkation Service ordered thirteen vessels of steel construction—nine large mine planters, each 172 feet long, and four river steamers, each 130 feet long. These were turned out by the quantity-production method known as "fabricated" ship construction; that is, the standard parts were manufactured in shops and then assembled into hulls on the launching ways. The contract for all thirteen went to the Fabricated Ship Corporation, of Milwaukee, Wisconsin, which created a new shipyard for the task. The company built parts in its own shops and let sub-contracts to manufactories located some miles away from the shipyard. Because plates came to the ways shaped and ready for the rivets, the system resulted not only in uniformity of construction but also in great speed. The contract for the so-called "junior" mine planters—eight boats, each 98 feet long—was taken by the Defoe Boat & Motor Works, of Bay City, Michigan, which also established a new yard.

When men first proposed to build iron ships, doubtless there were skeptics who derided the notion that safe and seaworthy vessels could ever be made of so heavy and sinkable a substance. When it was proposed to construct vessels of reinforced concrete, men could likewise be found who maintained that the "stone" boats would be, because of their ponderous weight, either unseaworthy or else unwieldy and uneconomical in operation. Although concrete hulls are indeed heavier than steel hulls of the same dimensions, the actual use of such vessels had demonstrated their practicability; and at a time when the nation's needs could scarcely endure any additional demand for steel, the utility of concrete construction for harbor craft was a veritable salvation.

The weight of a concrete boat, after all, is not so much greater than that of a steel boat of the same size as one would think. The concrete hull walls are made surprisingly thin. In most of the craft launched by the contractors of the Embarkation Service, the exterior concrete covering of the reinforcing bars did not exceed three-quarters of an inch in thickness. To mold concrete with such precision required, of course, constructive skill of the first order.

The Service concluded a contract with the Great Northern Shipbuilding Company, of Vancouver, Washington, for the construction of five concrete water-tank boats, each 100 feet long. To the Newport Shipbuilding Company went another contract to build nine concrete river vessels, each 130 feet long, at a new yard at Newbern, North Carolina. These river vessels were to be in the service of the Newport News Port of Embarkation. A third contract for six concrete car floats, each 265 feet long, went to the L. B. Harrison Shipyards (Inc.), at Athens, New York, and an identical contract to the Liberty Shipbuilding & Transportation Company of Cleveland, Ohio. Each one of these contractors set up new shipyards for the construction.

In appearance, the concrete boats differed little from steel craft of the same types. The service designers made, in the accepted curves and lines of hulls, no concessions to concrete. The river boats were built on the lines of standard steel ships, except that the sections amidships were practically square for about one-fourth of each vessel's length. The car floats were rectangular in shape, but curved slightly at the bilge and had raked ends. In fact, in all its ship construction the Embarkation Service considered the day when the Government might want to sell these vessels to private concerns. Every vessel was so designed that the possible private purchaser could some day use it in ordinary commerce and find no difficulty in placing insurance upon it.

The launching of a concrete vessel is a matter of careful engineering calculation. It is usually launched sideways, and great is its list before it regains equilibrium. The Embarkation Service was the first to launch large concrete vessels end on. Several of the car floats took the water in this way. Certain engineers who witnessed the first of these end-on launchings confidently expected to see the new float break in two and sink as soon as its bow struck the water; but the designers were sure of their calculations, and they awaited the result with confidence. The sagging or "hogging" stress which end-on launching puts upon the midship section of such a long and shallow-draft boat as a car float is tremendous. As the forward end of the hull goes into the water and is buoyed up by it, it is precisely as if the craft were picked up by one end on the sole support of the other end. In the steel of the first concrete car float launched endwise, the strainagraph, a special instrument for recording stresses in construction members, registered a stress of 4,000 pounds to the square inch. But this stress was well within the strength of the vessel, which neither buried its nose under water nor gave way amidships.

The Service pioneered elsewhere in concrete construction. It made the first adaptation of reinforcing to the inherent requirements of the hull. In previous concrete construction the reinforcing bars had been allowed to follow the converging lines of the hull and to come thickly together at bow and stern, thus actually making them stronger than the midship sections, where heavier strains occur. The marine architects of the Embarkation Service reduced the reinforcing at the ends of vessels and thereby made the hulls approximately 18 per cent lighter, with an actual gain in strength.

Pouring the concrete in marine construction is a delicate operation. The concrete forms or molds must be built for practically the entire hull before any concrete can go in; for after the pouring starts, it must continue until the entire hull is poured, to avoid the joints that result from pouring wet concrete upon hard. The shell of a concrete vessel is so thin that only 100-per-cent accuracy in the placing of forms will result in a successful hull. Extraordinary precautions must be taken to prevent the forms from spreading away from the pressure of the concrete as it is being poured in. This spreading the Service engineers avoided by running certain reinforcing bars of the deck through the side molds as tie rods. (The projecting ends were used later for fastening the wooden fenders on the exterior of the hull.) Nor could concrete be dumped in carelessly, as in ordinary construction. Precautions had to be taken

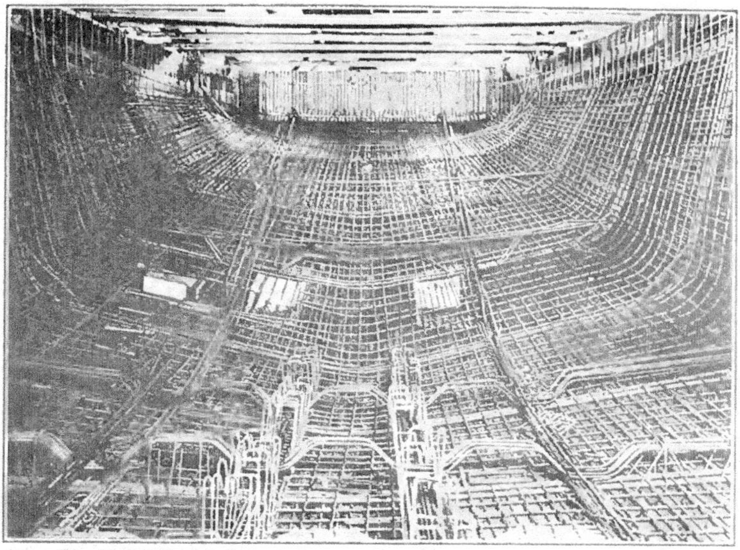

From The War College Collection

REINFORCING RODS LAID IN CONCRETE SHIP CONSTRUCTION

Photo by Signal Corps

KAPOK LIFE PRESERVERS, JACKET TYPE, SUPPLIED TO TRANSPORTS

Photo by Signal Corps

LIFE PRESERVERS ON *LEVIATHAN*

Photo by Signal Corps

LIFE RAFTS ON HOBOKEN ARMY PIER

to secure an absolute consistency of mixture, without interior cavities or holes. This consistency was obtained by battering against the forms with electric and pneumatic hammers, the jar packing the mixture and filling all interstices.

The Embarkation Service also designed a cargo barge of hollow tile and reinforced concrete. The hollow tile bottom, covered with concrete above and below, besides making the craft practically puncture-proof, allowed smooth concrete floors in the hold, upon which power shovels could operate in the unloading of coal and other bulk commodities.

One of the extensive activities at the New York Port of Embarkation was the supply and inspection of life-saving equipment for the troop transports. America sent across the Atlantic many ships loaded with men as ships had never been loaded before. The torpedoing of one of these vessels would have been a terrible disaster to the United States; yet the loss of life might not have been so great as the military authorities feared. To stack upon one of the great ships enough lifeboats to hold passengers and crew was physically impossible. But lifeboats, life rafts, and life preservers collectively provided floatage for every man aboard even the more heavily loaded transports. If a rescue came speedily after a sinking, most of the human cargo might have been saved.

The Government's sudden decision in April, 1918, to increase embarkation to the rate of 300,000 men a month taxed to the utmost the ability of the port utilities office, which had charge of life-saving equipment. The decision threw into our transport service an immense Allied merchant tonnage, and the War Department would not permit one of these foreign ships to sail with troops unless it possessed life-saving equipment up to the American army standard. The Government, instead of donating equipment to the British, French, and Italian vessel companies, sold it to them after the Port of Embarkation had procured it. During the period of overseas sailings the port utilities office bought thousands of life rafts and hundreds of thousands of life preservers.

The office also supervised the frequent cleansing of life

preservers. Amcriean troops were required by order to wear their life belts continuously night and day, sleeping and awake, while their transports were passing through the war zone—and the average transport was in it seventy-two hours. The Medieal Corps therefore insisted that the preservers be frequently washed and sterilized.

In the autumn of 1918, just after influenza had broken out in this country, a convoy carrying 46,000 men was about to sail for France. The port utilities office had approved the life-saving equipment. The life preservers were not dirty, but they had been worn on one previous voyage, since which they had not passed through the laundry. The convoy was to sail on Saturday. On the Thursday before, the Surgeon General condemned the complete equipment of life preservers on the convoy and demanded that sterilized equipment be substituted. The utilities office, in a hasty inventory of its stores, discovered only 4,000 life belts in stock. There was not enough time to clean the condemned life belts. The utilities officers began to comb the metropolitan manufacturing district for possible supplies of new life preservers which had been overlooked in previous purchases. In New Jersey they discovered a manufacturer who by mistake had made 25,000 kapok life belts more than he had needed to fill a previous government contract. The Port bought these belts outright and rushed all its available motor trucks to the factory in Newark. The trucking force freighted life preservers day and night. This find, together with the 4,000 life preservers on hand, brought the available supply up to nearly 30,000; but it still left a shortage of 16,000.

Then the utilities officers made by telephone a systematic canvass of the trade. They succeeded in picking up five hundred life belts here and a thousand there; but even after they had literally swept the trade clean, there was still a shortage. Meanwhile the office had been cleaning and sterilizing as many of the condemned preservers as it could; and, to swell the volume of this work, it temporarily chartered the plants of several large carpet-cleaning establishments in New York and operated them for two days and nights. The office was working in the sure knowledge that if it failed to provide the life preservers in time, the embarkation authorities would be forced to hold the convoy in port. No ship could sail without satisfactory emergency floatage for every man aboard. But the utilities office did not fail. The convoy sailed at the scheduled hour with complete life-saving equipment which had passed the inspection of the Surgeon General's department. To be sure, the utilities office had exceeded its authority in the emergency: only the Division of Purchase in Washington was empowered to buy life preservers. But the War Department granted, as in other like instances, *ex post facto* authority for the transaction. Thus the Port passed triumphantly through another crisis.

The Port of Embarkation maintained a disbursing office at New York to pay all embarking officers and men up to the date of their sailings. The payment was made, conveniently, in either francs or pounds, shillings, and pence, so that the soldiers would have no trouble with the money changers on the other side of the Atlantic. The Port disbursed millions of dollars in the payment of transient individuals and organizations.

The Navy operated most of the American troop carriers and incidentally supplied the food eaten by our soldiers on their voyages in such ships. Yet, by one of the queer inconsistencies of an organization so large and intricate as the Embarkation Service, the New York Port of Embarkation supplied food stores to several of the American transports throughout the heaviest period of travel and until June, 1919. The value of the stores so supplied was close to $3,000,000. The port utilities office took care of this work. It also operated the great port bakery in Brooklyn, and provided the food and the daily menus for the crews of all army harbor craft.

It required a special organization at the port to give the releases on which the Inland Traffic Service based its War Department Transportation Orders, the potent instruments, it will be remembered, that cleared away the great eastern freight car congestion in early 1918. This organization, the

Army Railway Traffic Service, was formed by the consolidation of the traffic offices which the several export bureaus of the War Department had maintained at New York. When the Army Traffic Service took hold, in the late winter of 1918, the Port could handle at the maximum 400 cars a day of export government freight. At the time of the armistice the Port was putting through an average of 1,800 carloads a day; the record day was 4,000 cars. The Service allocated space for freight on the cargo and passenger transports, stationed men throughout the New York yards to keep war department cars moving, handled the financial end of army shipping in the terminal, and worked its three office shifts twenty-four hours a day to keep up with the business.

Whenever commercial passenger vessels came into the transport fleet, it was necessary to refit them specially for such service. The work of refitting ships fell to the Port's vessel maintenance and repair division. All transports crossing the war zone were required to go armed. This meant the installation of gun platforms on both troop and cargo carriers. In equipping merchant passenger boats as troopships, the refitters literally crammed them with standee bunks, building these in wherever there were a few cubic feet of spare room, even constructing them in companionways which could be closed off. The constructors built rows upon rows of temporary staterooms in the transports, fitted up special bath houses on their decks, and installed ship hospitals. The hospital wards were entirely lined with white enamel, and their floors were of sanitary concrete composition. Each transport required at least one false deck on which to stow the additional lifeboats and life rafts required by regulations. The galley facilities of the ordinary passenger vessel had to be greatly expanded by the installation of ovens, steam tables, and other cooking equipment.

The Port's maintenance and repair division operated a large marine repair shop at the Hoboken piers to take care of the trip repairs needed by our transports. This shop was frequently of extraordinary usefulness. At one time during the cold weather of early 1918, the transport *Finland* was loading 2,000 troops for France. On the night before she was to sail, the entire sanitary system built on her decks froze tight, and many of the pipes burst. Under ordinary conditions the ship would have had to go to a yard for repairs and would have missed the convoy. The repair shop put her in condition within a few hours, and she sailed at the appointed time. During the period of embarkation the shop was manned with a military staff known as Ship Repair Shop Unit No. 301, an organization unique in the Army. The maintenance and repair division was later to render a most important service, for it was the agency which, after the armistice, converted the fleet of army cargo carriers into troop transports and thus enabled the Government to bring home the A. E. F. in so short a time.

From Hoboken departed the army couriers carrying urgent official mail from Washington to the general headquarters of the A. E. F. at Chaumont, France. For many months the War Department relied upon the regular mails; but the A. E. F. postal service labored under difficulties, and the dispatch of mail was slow. Several branches of the War Department sought to beat the mail by sending their more important letters to France by courier. Eventually the Embarkation Service undertook to perform this service for the entire Army. It set up the courier service, a tight little corps composed of a resolute handful of commissioned and non-commissioned officers. Courier headquarters were in Washington; a branch courier office at Hoboken attended to the details of courier travel. Back and forth between Chaumont and Washington sped the couriers, carrying sacks of mail in both directions and cutting down the delivery time by at least a week. The first courier started out in early July, 1918. From that time until June 30, 1919, the Service dispatched sixty-three couriers to France. A courier left Washington every fifth or sixth day. The average consignment of mail in charge of a single courier was thirteen sacks. In the first year of operation, the couriers took 614 sacks of mail to Europe and brought back 1,408 sacks.

As soon as this convenient service was inaugurated, some of the army producing bureaus took advantage of it to send to France urgently needed consignments of such small supplies as firing pins for machine guns and sighting instruments for field artillery. The accumulation of packages of this sort soon began to hamper the speed of the couriers; and presently the Embarkation Service created a kindred organization known as the War Department Overseas Express Service, whose traveling messengers handled official overseas parcels only.

The mail courier waited in Washington until the last train that would take him to New York in time to catch the convoy—usually a midnight train. He and his sergeant piled the mail sacks into a sleeping-car stateroom, locked the door, and stood watch and watch until morning; for nearly all the mail in their charge was highly confidential. At Manhattan Junction in the Jersey meadows a port motor truck was waiting next morning to transfer courier and baggage directly to Hoboken. If by chance the convoy had gone down the bay, the Port bundled the two soldiers and their mail sacks into a fast launch and hurried them to the rendezvous.

On the transport a room was assigned to the courier and his assistant. The two men lived with the mail all the way across the ocean. The sacks were piled in the stateroom, and one man or the other always watched them. Meanwhile Hoboken had cabled to Brest that a courier was on the way. The transport had scarcely lost headway in the French harbor when a launch darted out from shore. Into this the courier and his assistant loaded their sacks. Often they were on the Paris train, locked in a special compartment, before any of the troops on their ship had yet landed. At Paris an A. E. F. courier met the train to take over the sacks directed to our army representatives in England. The American courier continued on to Chaumont with the G. H. Q. mail. The swiftest trip between Washington and Chaumont was made in six days and a few hours. The courier who set this record left Washington at midnight, early the next morning boarded a ship in a 20-knot convoy, barely caught the Paris train at Brest, and made a perfect train connection in Paris.

The courier service made itself extremely valuable by taking to France the cargo manifests of freight transports ahead of the arrival of the ships. It was out of the question to send such voluminous itemized invoices by cable; yet it was important that the army ports in France receive the manifests before the cargo arrived. The regular post office could not be relied upon to forward the manifests in time. Therefore, as soon as the American ports dispatched cargo transports they sent the freight manifests to Washington, and the Embarkation Service forwarded them to France by the first courier. The couriers also brought to the United States the lists of American casualties in the war.

When the Peace Conference opened in Paris, the State Department required almost exclusive use of the Atlantic cables, and the use of the cables by the Army was cut down to little or nothing. The War Department then turned to the next fastest means of communication, the couriers, and invented a new kind of official dispatch, known as the courier cablegram. The couriers carried much mail between Washington and the Peace Conference. Each eastbound courier had at least one pouch of state department mail for the American Peace Commission, and some of the returning couriers carried as many as six pouches addressed to the State Department. All the mail between the White House in Washington and President Wilson in Paris was carried by army couriers. During the period of hostilities the dozen or so couriers were the only soldiers attached to a home station who were permitted to wear cloth shoulder insignia. After the armistice, the New York Port of Embarkation adopted a shoulder insigne, a gray cloth rectangle upon which was worked the monogram P. of E., N. Y. This and the courier badge were the only home insignia. Both embarkation-service and A. E. F. couriers wore on their shoulders the silver greyhound, a fit emblem of their service.

Couriers were favorite children of the Army. Quick passages

by the American couriers brought rewards in France, in the form of trips to European cities in the A. E. F. courier service. After the armistice some of the American couriers went by airplane from Paris to Brussels, even to Rome; others traveled by train to the Balkans and to Constantinople. The little band developed a proud *esprit de corps*, of which one evidence was its pert self-applied nickname, "Fast Company."

Photo by Signal Corps

EMERGENCY LIFE RAFTS FOR TROOP TRANSPORTS

Photo by Signal Corps

EQUIPPING LIFEBOATS FOR TRANSPORTS

Photo by Signal Corps

EMBARKING TROOPS MARCHING THROUGH NEWPORT NEWS

Photo by Signal Corps

TROOPS APPROACHING PIER, NEWPORT NEWS

CHAPTER XXI

AT NEWPORT NEWS

HOBOKEN possessed no monopoly of the transoceanic shipment of soldiers and military supplies. On the shores surrounding Hampton Roads in Virginia there was another large enterprise in transportation, conducted by a port of embarkation equal in rank to that at New York, with its own command and its own complete installation of utilities. New York specialized in the embarkation of troops. The Newport News Port of Embarkation sent out some hundreds of thousands of troops, but its specialty was the export of supplies, particularly heavy supplies, to the A. E. F.—notably steel rails, motor trucks, ammunition, and high explosives. The physical equipment of the Newport News Port of Embarkation included such installations as the General Ordnance Supply Depot at Pig Point on the southern shore of Hampton Roads, the Norfolk Engineering Depot, and the great Norfolk Army Base. These were among the largest establishments in the whole army supply chain. Also at Newport News was the principal embarkation depot for the horses and mules sent to the A. E. F.

As a sender of troops to France, Newport News tended to specialization. It was the principal embarkation point for stevedores and labor troops; also, many of the artillery regiments boarded ship there. It embarked balloon companies and airplane squadrons. The first embarkation at Newport News occurred on January 17, 1918, when eight aëro squadrons sailed on the transport *H. R. Mallory*. From that time on, various scattering units sailed at intervals until April, when the troop embarkation from Newport News began in earnest.

At the time of the armistice Newport News had sent overseas nearly 300,000 troops.

This record best illustrates the significance of the Virginia port when it is compared with what occurred at other Atlantic seaports. The earliest embarkations of all occurred at New York in the spring and summer of 1917. The first embarkation of American troops from any port other than New York took place on September 16, 1917, when the 102d Infantry with a field hospital and an ambulance company, all units of the Twenty-sixth Division, boarded the White Star Steamship *Canada* at Montreal. The next outside embarkation occurred at Philadelphia October 16, 1917, when the 3d Cavalry Regiment sailed on the S. S. *Northland*. On the day before Christmas, 1917, the White Star liner *Canada*, which had carried the first American troops from Montreal, took on board the 24th Machine Gun Battalion, a trench-mortar battery, and an evacuation hospital, at Portland, Maine. On April 13, 1918, Boston began functioning as an embarkation sub-port, the 153d Infantry, the 306th Infantry, and the latter's machine gun battalion embarking there on the Cunarder *Karoa*. On May 26, 1918, occurred the first embarkation from Baltimore, Maryland, the 303d Engineers boarding the S. S. *Ajax* of the Blue Funnel Line. Now, Montreal and Quebec (during the winter, when the St. Lawrence River was not navigable, the ports of Halifax, Nova Scotia, and St. Johns, New Brunswick, were used by our Embarkation Service instead of Montreal and Quebec), Philadelphia, Portland, Boston, and Baltimore ranked in the military organization as sub-ports of the New York Port of Embarkation, their activities being commanded from the headquarters at Hoboken; but Newport News possessed no auxiliary ports.

At New York there embarked 1,656,000 troops, at Newport News 288,000, at Boston 46,000, at Philadelphia 35,000, at Montreal 34,000, at Quebec 11,000, at Portland 6,000, at Halifax 5,000, at Baltimore 4,000, and at St. Johns 1,000. These (approximate) figures fairly measure the activities of these various ports as points of embarkation for troops.

The work of the Newport News Port of Embarkation was similar to that of New York, though it was on a smaller scale. Newport News had its embarkation camps and its inspection and supply services. Units ordered overseas via Newport News went through much the same embarkation mill as that at New York. The problem of handling troops at Newport News was never exacting; but for all that the Port had its own peculiar troubles. It began its existence with port facilities much farther from adequacy than those which the Army was able to utilize immediately at New York. The whole plant at Newport News had to be built from the ground up, and while the construction was in actual progress the Port was called upon to handle an immense quantity of supplies, especially engineering materials needed in France ahead of the arrival of the main body of troops. The history of the Port at Newport News is, then, largely the history of a struggle to overcome difficulties, provide a working equipment, and whip into shape a smoothly working organization.

The ground over which the Port held jurisdiction is historic. Here first set foot on the soil of this continent some of the earliest English settlers sent out from the mother country to America. Close to Newport News are the ruins of the abandoned town of Jamestown, Virginia, the first English settlement on the Atlantic coast. In this region the Indian princess Pocahontas, daughter of Chief Powhatan, saved the life of Captain John Smith and, later, married John Rolfe, the wealthy English merchant. It is worth noting that two of the army transports which sailed regularly from Newport News were the *Powhatan* and the *Pocahontas*, both formerly German liners, remodeled to carry thousands of American soldiers to the battle fields of France. One of the first settlements in this region was planted by Captain Newport, who commanded a small squadron of colonist ships which he brought safely into Chesapeake Bay in the spring of 1607. The city of Newport News, however, was named, not for this pioneer, but for Sir William Newce, an Irish gentleman who sponsored the voyage of Master Cookin in 1621, in a vessel carrying eighty colonists. These pioneers settled on the James River and named their settlement New Port Newce, which the phonetic tendency in spelling later modified to its present form.

The Government selected Newport News to be a port of embarkation, not because of its historical associations, but rather because of the splendid deep-water facilities of the great inland harbor known as Hampton Roads. This salt-water basin is not only large enough to give safe anchorage to the greatest vessels afloat—it has accommodated the whole North Atlantic Fleet at once—but it is also doubly land-locked. Hampton Roads itself has only a narrow entrance into Chesapeake Bay, which in turn is enclosed from the ocean by the Virginia capes. The whole Chesapeake was securely guarded against possible enemy attack by the forts on Cape Charles and Cape Henry and by submarine nets stretched from one cape to the other. Fortress Monroe, commanding the inlet to Hampton Roads, gave additional protection to the vital military establishments to be built along the shores of that body of water. The northern shore of Hampton Roads is the end of the long peninsula enclosed between the York and the James rivers, which in this region are not so much flowing streams as broad, deep branches of the sea. On the southernmost point of this peninsula is the city of Newport News, facing south across Hampton Roads toward the city of Norfolk, which is located at the extreme southern corner of the bay. The facilities built for the Port of Embarkation almost surrounded this deep and sheltered body of water.

At the beginning of the war, when it became certain that the Army would establish a port of embarkation at Hampton Roads, there was a sharp contest between Norfolk and Newport News over the question of which city should be the port headquarters. Both possessed advantages. On each side of Hampton Roads were located some of the most modern coaling docks in the world, at the tidewater terminals of railroads leading from the coal fields of West Virginia. The pier of the Chesapeake & Ohio Railroad at Newport News could load 6,600 tons of coal an hour. Both cities offered extensive ship-

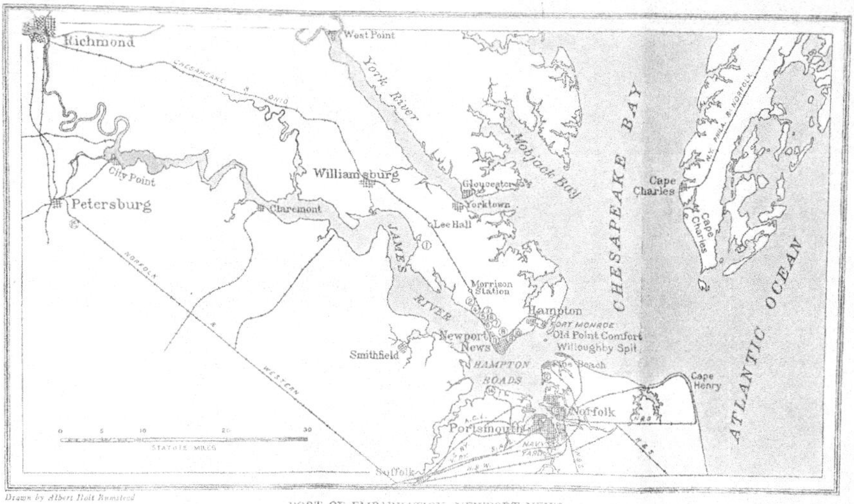

Drawn by Albert Holt Rumstead

PORT OF EMBARKATION, NEWPORT NEWS

LEGEND
1. Camp Abraham Eustis
2. Camp Morrison
3. Camp Alexander
4. Animal Embarkation Depot No. 301
5. Camp Hill
6. Open Storage
7. Warehouses
8. General Quartermaster Supply Depot
9. Camp Stewart
10. Debarkation Hospital
11. Port Administration Headquarters
12. Norfolk Army Base
13. Norfolk Engineer Depot
14. Pig Point Ordnance Depot
15. Camp Lee

repairing and drydocking facilities. Norfolk was near the Portsmouth Navy Yard; at Newport News was located the Newport News Shipbuilding & Drydock Company, operating one of the largest shipyards in the world. An investigation by an army board in the spring of 1917 made it evident that neither city could accommodate all the utilities of a complete port of embarkation. The board recommended the use of sites in both cities and at other points on the shores of Hampton Roads. Since it was convenient, because of railroad and other conditions, to establish most of the port fixtures on the Newport News side, that city was selected for the headquarters. Major General Grote Hutchinson, Commander of the Port, set up his headquarters at Newport News on July 11, 1917.

Construction was a much more dominant characteristic of the war activities at Newport News than it was at Hoboken. The entire period of hostilities was marked by continuous expansion in military building on the shores of Hampton Roads. The vicinity became the site of more war establishments than were located in any other equal area in the United States. There were great aviation fields and divisional and special training camps near by; there was also the principal overseas supply base of the Navy, as well as a great naval training station. The Port itself soon developed into an enormous and constantly growing institution, its growth being ended only by the armistice.

Like New York, the Newport News Port of Embarkation possessed two principal embarkation camps. These were Camp Hill and Camp Stewart. Camp Stewart, the larger, was located on the shore of Hampton Roads within the eastern city limits of Newport News. Its site was an area 215 acres in extent; and the finished camp had, in the section given over to transient troops, capacity for 450 officers and 15,600 enlisted men. The camp was operated by a permanent force of 40 officers and 1,230 enlisted men, troops of the 48th Infantry Regiment and the 11th Cavalry.

Adjoining Camp Stewart on the east and occupying a cool and pleasant location with a beautiful outlook across Hampton Roads, was the port debarkation hospital. A few of the hospital wards were open in the autumn of 1917. Construction continued, and a year later the hospital could accommodate nearly 4,500 patients. It had become one of the largest institutions of its sort in the world. Here were landed thousands of the American soldiers who had been wounded in the fighting in France.

Camp Hill was located on the shore of the James River, about a half mile north of the city limits of Newport News. This was a smaller camp, with accommodations for about 350 officers and 6,900 enlisted men.

Three other embarkation camps for special sorts of troops became integral with the port plant. One of these was Camp Abraham Eustis, located some twelve miles up the James River on the Chesapeake & Ohio Railroad, not far from the army balloon school at Lee Hall, Virginia. Camp Eustis was a combination training and embarkation camp for heavy artillery regiments. It possessed a target range where big guns could be fired under the direction of observers in captive balloons; so that the target practice gave training to both the artillerymen and the aërial observers from Lee Hall.

Camp Morrison was the Air Service's general supply depot and embarkation camp at Newport News. It was also located on the shore of the James River north of Newport News, about half way out to Camp Eustis. The camp site was a mile-long narrow rectangle containing 300 acres. The C. & O. Railroad supplied good transportation facilities between the camp and the embarkation piers at Newport News. The camp equipment included twenty-four warehouses for the storage of aëronautical equipment, and also twenty-seven barrack buildings for Air-Service troops. To this point came many of the Army's aëro squadrons, where they were inspected and equipped for overseas service and, in some cases, given additional training while waiting for their transports.

Camp Alexander, the third of these attached camps, was a special camp for the training and embarkation of stevedore regiments and labor battalions. Here were trained and housed, also, the permanent labor detachments assigned to the Port of Embarkation at Newport News. Camp Alexander was located just east of Camp Hill.

One of the most important camps at Newport News was the animal embarkation depot, known in the army organization tables as Animal Embarkation Depot No. 301, situated between Camp Morrison and Camp Hill. The depot was one of the first utilities built at the port. Its corrals, fenced fields, and pens gave accommodation to a maximum of 10,000 animals.* The animal embarkation depot occupied seventy-seven acres. At the railroad tracks were six unloading pens, each of capacity for 350 horses or mules. The unloading pens also served as detention corrals in which animals might be quarantined to determine the presence or absence of disease. Behind the unloading pens were twenty-four regular corrals, each for 350 animals. The depot maintained a great animal hospital for the treatment of the various maladies which afflict beasts of burden. The hospital equipment included baths and tanks for the disinfection of animals before their admittance to the regular corrals.

Practically all the animals sent to the A. E. F. passed through this depot. Up to the end of March, 1918, the embarkations of animals from Newport News had approached 30,000. At that time came the urgent call for troops in France, an event which made it necessary to divert all ocean cargo tonnage to the shipment of supplies; and the War Department ceased altogether to send animals to France. On August 12, 1918, animal shipments began again; and between that date and the end of November nearly 40,000 horses and mules left the United States for France, most of which embarked at Newport News.

All the rest of the port establishment had to do with the

* In an emergency the Army had the use of corral accommodations for 5,000 more at the British remount depot near by. Nearly all of the horses and mules purchased by the British Army and shipped overseas prior to 1917 passed through this remount depot at Newport News. The British began shipping American animals in December, 1914. Our first shipment occurred on October 14, 1917.

shipment of army supplies. The equipment included a line of seaport warehouses in the city of Newport News, providing 450,000 square feet of storage for supplies awaiting loading. These warehouses, each one story high, were plentifully interspersed with fire walls to prevent any considerable conflagration. In spite of the dangerous composition of many of the materials handled through Newport News, there was no fire or explosion of any consequence during the whole overseas movement.

In the city of Newport News was also located a general depot for the storage and issue of quartermaster supplies to embarking troops at the many training camps in the vicinity. This depot occupied a large abandoned brewery near Camp Stewart, and it administered in addition the quartermaster warehouses in the various embarkation camps.

The administrative headquarters of the Port occupied a temporary two-story building which covered an entire city block in Newport News.

While troops and animals were embarking at the city of Newport News, the Port conducted its great supply export business from establishments that dotted the eastern and southern shores of Hampton Roads. One of the largest of these was the huge army base on the eastern shore of Hampton Roads, a few miles north of Norfolk. This installation, like the Brooklyn Army Base, was of permanent concrete construction. It consisted of great warehouses and covered piers for the handling of general supplies for the A. E. F.

South of the Norfolk Army Base, within the northern corporate limits of Norfolk, was the General Engineer Depot of the Army. The physical plant the Army found ready for occupancy when war was declared: it had just been built by the Norfolk & Western Railroad for its own uses. With the improvements and conveniences added, this was one of the most complete ocean terminals in America. At its two covered piers eight ocean vessels could be tied up for simultaneous loading. The Army built additional warehouses and installed large stationary cranes and other loading machinery for han-

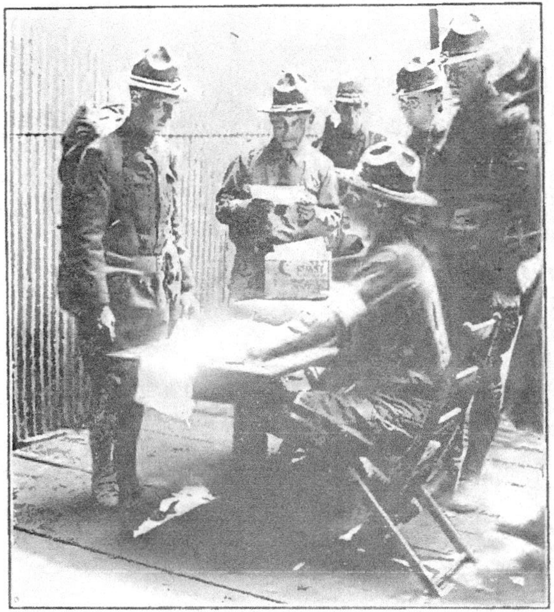

Photo by Signal Corps

CHECKING TROOPS ABOARD TRANSPORT, NEWPORT NEWS

Photo by Signal Corps

LOADED TROOPSHIP LEAVING NEWPORT NEWS

Photo by Signal Corps

BOARDING SHIP AT NEWPORT NEWS

Photo by Signal Corps

CROWDED TRANSPORT LEAVING PIER, NEWPORT NEWS

dling heavy articles. Much of the engineering material sent to the A. E. F. was such heavy bulk freight as steel rails; and the Engineer Corps found it necessary to lay out at Norfolk a classification yard, in which freight of similar sorts, coming in by rail, could be dumped together in piles, upon which cargo ships could draw for complete homogeneous loads. For this purpose the Army leased 170 acres at Portlock, Virginia, just south of Norfolk.

On a site of 600 acres on the south shore of Hampton Roads, a few miles northwest of Portsmouth and, by water, five miles south of Newport News, the Army built the huge Pig Point General Ordnance Supply Depot. The construction of this institution began in November, 1917, and was not complete a year later, although by that time the depot had a tremendous capacity. Its principal function was to store and export ammunition and high explosives. Since the water along the southern shore of Hampton Roads is comparatively shallow, all loading from the ordnance depot had to be by lighterage. The Port set aside a definite area of Hampton Roads near Pig Point for the anchorage of powder and ammunition ships, so that in the event of an explosion the damage to the port plant would be confined to a minimum. Except tugs and lighters operating between the powder ships and the ordnance depot, no vessels were allowed to enter this area. The depot consisted principally of chains of powder and ammunition magazines. Its site was protected by an unscalable wire fence with sentry-boxes every 500 feet along its four miles of length. At the time of the armistice the depot could receive and export a hundred carloads of powder and ammunition every day.

The Government conducted at Hampton Roads a great housing enterprise which provided living quarters for thousands of men in the shipyards, the ship-repair yards, and the construction gangs. Many barracks were built for the troops stationed at the port. The project also included a temporary hotel in Newport News for casual officers awaiting embarkation.

Early in the war, A. E. F. supplies flooded in upon Newport News, and for many months the port organization was primarily engaged in getting its head above water. But gradually the port facilities began to measure up to the requirements, and in the end Newport News could meet well-nigh any conceivable demand. The first vessel to be loaded there was the *Momus*, which sailed September 3, 1917, with a cargo which included 1,605 bales of hay and 80 dismounted motor trucks. (The first freight convoy from Chesapeake Bay assembled in October, 1917, but none of its ships was loaded at Newport News.) The first animal ship was the *Amphion*, formerly the German *Köln*, which sailed October 14 with 881 mules, 169 horses, and forage and general cargo.

The usual procedure for a cargo vessel at Newport News was to visit, first, the piers at the engineer depot at Norfolk, where it took on heavy materials for ballast, and then to proceed to the army base or to the piers at Newport News for general supplies.

The Port repaired ships and refitted ships for transport service in the yards of the Newport News Shipbuilding & Drydock Company. This was not a felicitous arrangement, for the facilities of the shipbuilding company were sorely needed by other government agencies. Had the war continued, the Newport News port equipment would probably have included a large ship-repair yard. The Port produced animal transports by building stalls into cargo vessels, took ships from the tropical trades and built on deck protection so that they could brave the winter weather conditions of the North Atlantic, and also refitted for the military service many of the ex-German vessels—particularly ex-German cargo ships, some of which had been seized by Cuba, Brazil, and other Latin-American co-belligerents.

Newport News conducted a large salvage enterprise for reclaiming food, clothing, and other supplies left behind by embarking troops. The Port also handled a few casuals. These fell under the command of a special officer, who compiled the service records of such soldiers and organized them into companies. Finally, Newport News, being the Army's principal cargo port, was confronted by the necessity of providing crews for the cargo carriers of the army fleet—a difficult matter, because the Norfolk district was a poor labor market. Labor scarcity of all sorts was a hampering factor throughout the entire development of Hampton Roads. Eventually it became next to impossible to secure civilian crews for our cargo transports without delaying them in port. This difficulty virtually disappeared as, more and more completely, the Navy took over the operation of cargo vessels and manned them with crews of enlisted men.

Made in the USA
Monee, IL
04 November 2026

67387523R00223